under the laws of the State

named, under the title

way and Light Company,

Revised Statutes of said Sta

. Said Articles of Incorpo

eting and read at length

d unanimously

Articles of Incorporation of

and Light Company, org

of Wisconsin on January

Path of a Pioneer

A Centennial History of the Wisconsin Electric Power Company

by John Gurda

Library of Congress Catalog Number: 96–60703

ISBN: 0–9652461–0–8

Contents

Imagine, if you can, a world without electricity. Unplug the computer. Turn off the lights. Put away the TV and the toaster, the CD player and the microwave. Fall back to a world of kerosene lamps, steam engines, and horse-drawn streetcars. Revisit, without nostalgia, the darkness and isolation of our nineteenth-century ancestors.

Of all the marvels that make modern life possible, electricity is the one we take most for granted. The light on the porch, the image on the screen, the hum of the refrigerator are so much a part of our daily routines that they practically vanish into the background. But it is self-evident that our way of life rests on a foundation of electric power. Before the telephone, before motion pictures, before the radio or the television or the computer, there had to be a dependable source of current. As water is to life itself, electricity is to the life of modern society.

To more than two million people in Wisconsin and Upper Michigan, electricity is the Wisconsin Electric Power Company. The average customer rarely looks beyond last month's bill, but the company behind the bills has an uncommon character, a distinctive identity rooted in a remarkable history. Wisconsin Electric was created by an Eastern capitalist who saw Milwaukee as the cornerstone of a national utility empire. The firm was carried to maturity by a Midwestern autocrat hell-bent on providing the best electrified transit service in the country. Its emphasis shifted from traction to power under the direction of a 31-year-old whiz kid.

As the twentieth century progressed, the company weathered power shortages in the World War I era, bitter turf wars in the 1920s, economic collapse in the 1930s, and an unprecedented demand for energy during World War II. When the long postwar boom fizzled in the 1960s and '70s, Wisconsin Electric responded with trailblazing programs in conservation, communication, and economic development. When barriers to free competition began to fall in the 1990s, the company broke new ground again, choosing a path that has led to the brink of a merger with a neighboring utility.

Wisconsin Electric's history has been, in a word, tumultuous. The company emerged from conditions of absolute chaos. Internal power struggles kept it off-balance even after the utility was well-established. Conflicts with external authorities added a constant political overtone to all its endeavors. The utility went through multiple incarnations as it evolved, and its course was shaped decisively by public controversies, technical breakthroughs, and economic travails. Pausing at the century mark, Wisconsin Electric looks back on a heritage that is by turns edifying, surprising, and occasionally outrageous.

That heritage is also complicated. The business itself was unusually complex; Wisconsin Electric's ancestors worked simultaneously in two vastly different fields — electricity and transportation — and the company's "family tree" is a Byzantine maze of mergers and acquisitions, branchings and prunings. Today's utility evolved from more than a dozen major enterprises and literally hundreds of smaller ones. The company's multiple layers and dimensions, particularly those of its first half-century, seem to defy coherent presentation in story form. Although every attempt has been made to give this narrative an orderly chronological flow, readers will find themselves learning about streetcar routes, rural electrification, and political crusades in the same chapters.

For all its complications, the centerline of the story is clear: Wisconsin Electric has been a pioneering agent of pervasive change. Power would have come in any case; streetcars would have clattered through downtown Milwaukee, and lights would have flickered on from Iron Mountain to Kenosha. But there was nothing inevitable about the course that power took in the company's service area. Responding to unique circumstances and reflecting the force of unique personalities, Wisconsin Electric generated the region's future. A century after its founding, the company looks ahead to even broader tomorrows.

Chapter 1

1880-1890

The Coming of a Second Fire

It was only a matter of time. After 1831, when Michael Faraday, for one, developed a working generator, it was only a matter of time before electricity transformed the world. Faraday's device — a copper disk rotating between the poles of a horseshoe magnet — was deceptively simple. Although it was little more than a science-fair experiment by modern standards, his generator harnessed the power that drives the modern age. A primal force of nature had been uncovered at its root; lightning had been reduced to human dimensions and placed under human dominion. Electricity soon crossed the threshold from the scientist's laboratory to the inventor's workshop, and nothing was ever the same again.

Nearly 50 years would pass before Faraday's discovery revolutionized life in Wisconsin. Progress on the local level awaited the development of an entire industry, one whose shape was determined by a fractious group of inventors, promoters and, surprisingly, politicians. Primordial chaos reigned in the electrical industry's formative years. From that chaos emerged Wisconsin Electric Power Company, the local agent of a global revolution.

Inventors played the crucial early role. Their initial challenge was not to make electrification possible, but to make it practical. A small army of tinkerers, most with a profit motive, pushed back the technical frontiers, and the names of a few became household words. Samuel Fairbanks Morse invented the telegraph in 1837, and Alexander Graham Bell developed its younger cousin, the telephone, in 1876. Both devices used electrical pulses to communicate, one in code, the other in a simulation of the human voice. Other inventors considered these low-power applications only useful beginnings. A number of visionaries came to see electricity as nothing less than a second fire, a new source of light, warmth, and power that would prove just as indispensable as the first flames that flickered against the walls of Neolithic caves.

The early wizards of the workshop followed two major paths: lighting and transportation. Those paths would ultimately converge, providing a round-the-clock load for America's central stations, but they were, in the beginning, as separate as night and day. In 1876 Charles Brush, a full-time chemist and part-time inventor based in Cleveland, developed a commercial version of the arc light. The Brush lamp generated, in essence, continuous man-made lightning. An electrical current jumped a precisely controlled gap between two carbon rods, producing a flickering light so bright it was called an "artificial sun."

Thomas Edison, a figure of mythical proportions in American history,

Mounted on spindly steel towers, arc lights turned night to day in hundreds of American cities.

worked on the other end of the scale. In 1879 he brought the sun indoors, developing the world's first practical incandescent light. Just as important, Edison devised a complete system — dynamos, conduits, fuses, insulators, and all the rest — that brought the light to the public. On September 4, 1882, that system came to life with the opening of the celebrated Pearl Street Station in lower Manhattan.

Frank Sprague played a comparable role on the transit side. Street railways powered by horses and mules had been fixtures on the American scene since well before the Civil War, but the systems began to change in 1887, when Sprague electrified a line covering 12 hilly miles in Richmond, Virginia. "Lincoln set the slaves free," proclaimed his backers, "and Sprague has set the mule free."

Brush, Edison, and Sprague were hardly alone. In the 1880s, a pivotal decade in the history of technology, hundreds of inventors-turned-entrepreneurs challenged the early leaders for a share of the market, and the result was competitive chaos. America's electrification was every bit as disorderly as an Oklahoma land rush, where success in a crowded field depended on the speed of your horse and your ability to elbow out competitors. Spurious claims and libelous counterclaims, greased palms and cutthroat deals were the order of the day.

The chaos was especially acute in the nation's cities, which provided the largest natural markets for electrical systems. Unlike merchants or manufacturers, street-lighting and streetcar firms could not simply set up their businesses and solicit customers. Because transmission lines and trolley

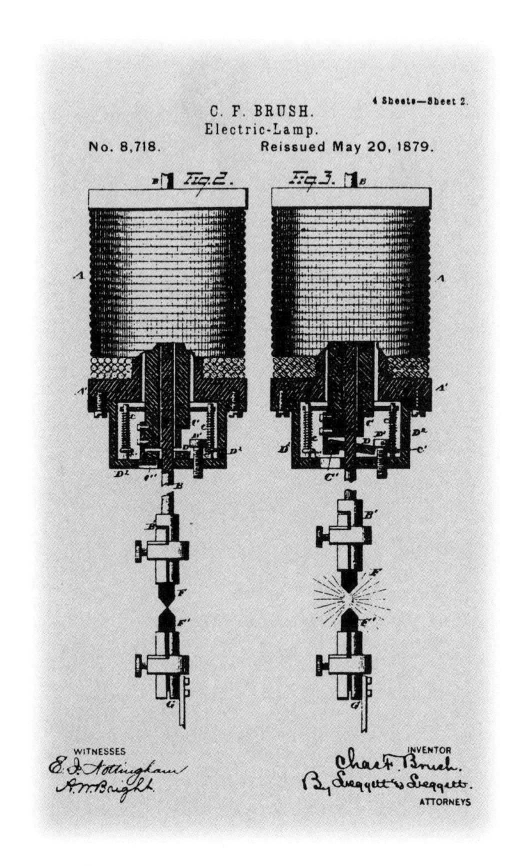

One of several arc-light devices patented by inventor Charles Brush

tracks used public ways, both were subject to regulation by public authorities. Electrical firms could not begin without municipal franchises, and they could not continue without meeting conditions set by municipal officials. Politics, therefore, intruded early and often, adding another layer of complexity to an already chaotic situation.

Both the competitive climate and the political atmosphere reflected a fundamental change in the nation's weather. The American way of life was transformed between 1865 and 1900, a period that Mark Twain dubbed the Gilded Age. A rural agrarian society became an urban industrial society, and the country's pace quickened dramatically. The spirit of the age was wildly expansionist; immigration, the westward movement, and international adventures enlarged both the dimensions of the country and the ambitions of its people. In a time of prolonged peace, ambition blossomed into avarice, and the nation came to share a fervent belief in material progress, whatever its cost in dollars or ideals. The easy virtue of the Gilded Age made business and politics practically indistinguishable. Not surprisingly, two representative figures of the era were the robber baron and the political boss.

Both types played leading roles in the creation of what is now Wisconsin Electric. The company's early years are, in fact, a microcosm of the American story of electrification. With the foregoing developments as context, Chapter 1 chronicles the coming of a second fire to eastern Wisconsin. Complex, colorful, and sometimes scandalous, the story breaks neatly into two halves — lighting and transportation — that form a whole chapter. In the formative years of the electrical industry, lighting and transit were as separate but parallel as a pair of railroad tracks. It was not until 1890, after years of chaotic competition and political intrigue, that the tracks began to converge in a unified regional system.

I. Turning on the Lights

The story of electrification follows a definite sequence regardless of locality. First came the scientists, men like Michael Faraday in England and Joseph Henry in the United States, who demonstrated the basic principles of electromagnetism. They were followed closely by the inventor/entrepreneurs — Brush, Edison, Sprague, and numerous others — who moved electricity from the realm of theory to the real world of applications. Finally came the nomadic army of sales agents who brought power and light to the masses. Many were tinkerers in their spare time, and all approached their task with missionary zeal. They were, in effect, itinerant preachers, bringing the gospel of electricity to cities and towns across the country.

Henry Evans Jacobs was the first electric circuit rider in Wisconsin, and ultimately the most important. His influence was so pervasive that historian Forrest McDonald called Jacobs "the godfather of electric service in Wisconsin." Born in Chillicothe, Ohio, and raised in Toledo, Henry Jacobs migrated to Chicago in about 1880. He soon landed a job as an arc lighting agent for Brush Electric, a firm that was only months old. Although his territory covered the entire upper Midwest, Jacobs concentrated his efforts on Wisconsin, selling both private installations and central stations.

(The difference between the two systems is subtle but important. The same equipment powered both, but private installations existed for private use on private property — halls, stores, factories, and other large spaces. Central stations, by contrast, were open to all comers in cities and towns; they were public utilities subject to public regulation. The difference between the two systems is like the distinction between a private automobile and a taxicab.)

In 1881 Henry Jacobs made his first sale in Wisconsin, a private installation that lit the Wisconsin Central Railroad's shops in Stevens Point. It was the first commercial arc-light system in the state. In January, 1882, Jacobs established, with local investors, the first central station in Wisconsin, a small arc-light firm serving stores in downtown La Crosse. The agent then moved on to Fond du Lac, an industrial center at the foot of Lake Winnebago. In June, 1882, after recruiting a group of local backers, he secured a city contract to build five towers at strategic points around the town, each bearing four 2,000-candlepower Brush lights. The system, promised a local newspaper, "will be adequate to light every part of the city as brightly as good moonlight." Some of Fond du Lac's leaders had been advocates of arc lighting from the day the systems became available, but the city had an extra incentive: Fond du Lac was the site of the State Fair in 1882. Over the deeply felt objections of the local gaslight company, the Brush system was rushed to completion, and the lights went on during the evening hours of September 8 — one day before the fair opened.

Jacobs practically lived in Fond du Lac during the construction phase. Like most early manufacturers, Brush Electric had a major stock interest in all its local affiliates, and the salesman was responsible for protecting Brush's Fond du Lac investment. He also found time to fall in love. Jacobs married Martha Dodd, daughter of the local railroad express agent, in December of 1882. Fond du Lac became his second home, and he retained close ties there long after the lights went on.

The enterprise Jacobs sired was, like so many pioneer utilities, something less than a runaway success. Wind and lightning practically destroyed the system more than once, and its eco-

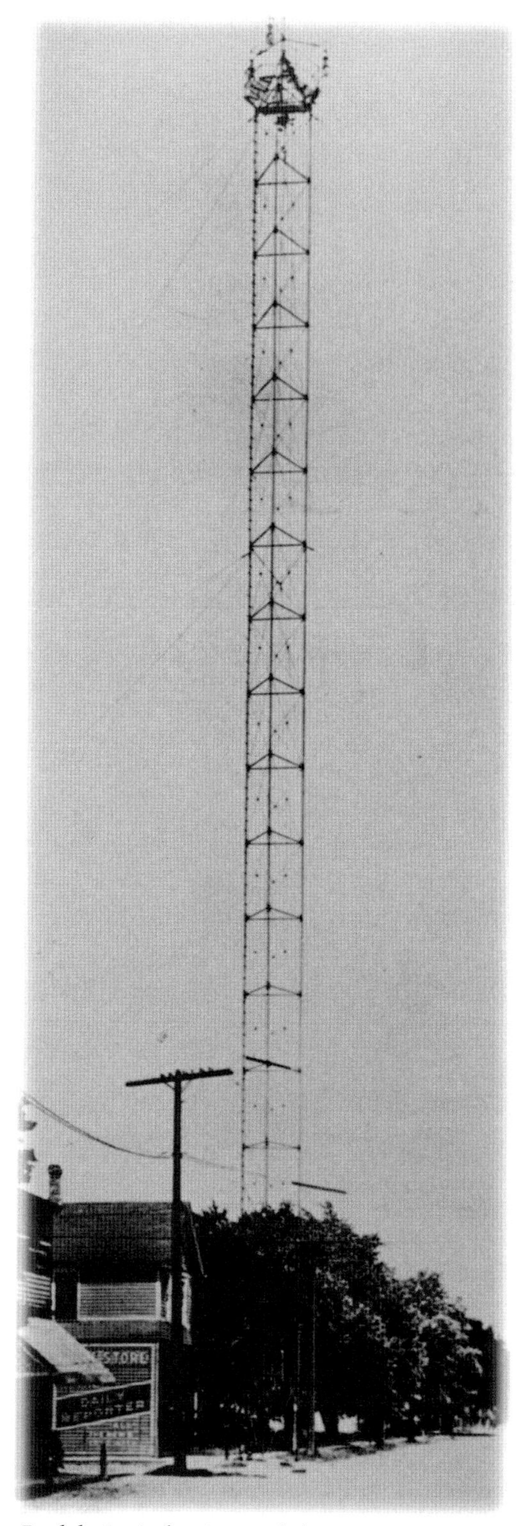

Fond du Lac's electric street-lighting system, the first in the state, consisted of five towers like this one mounted off Main Street.

nomic history was just as stormy. But the Fond du Lac installation is remembered as the first electric street-lighting system in the state.

Jacobs' next sale made history on an even larger scale. On May 25, 1882, months before the Pearl Street Station opened, the Edison interests had established the Western Edison Electric Light Company in Chicago. Serving Illinois, Wisconsin, and Iowa, the subsidiary was organized to sell the "isolated" incandescent plants that Edison was already producing in quantity. Although he did not sever ties with Brush Electric until the Fond du Lac system was completed, Henry Jacobs saw a brighter future in the incandescent light. He became Western Edison's Wisconsin agent within days of the company's incorporation.

The new disciple's first stop was Appleton, a city of 10,000 on the banks of the Fox River. Appleton had begun as a college town — Lawrence University was chartered in 1847 — but entrepreneurs had made it a mill town as well. The river dropped sharply in the vicinity of Appleton, making it an ideal source of industrial power — enough power, claimed a local booster, to provide employment for 250,000 factory hands. By the 1880s, "the flats" along the river bottom boasted a network of dams and canals, and "the water power" was turning the wheels for a number of plants, particularly paper mills. Appleton was booming, and Henry Jacobs thought the town was ready for the electric light.

Ever mindful of the need for publicity, he paid an early visit to the office of the *Appleton Post*. The newspaper's editor poked a little fun at his caller in the edition of July 20, 1882:

> *There are two or three objections to this new-fangled light which Mr. Jacobs never alludes to. One is that when the streets are lighted on the Edisonian plan, deacons of churches will not be able to emerge from corner groceries wiping their lips, and young fellows will not be able to hug fair demoiselles in the shadow of the trees. Personally Jacobs is not a bad sort of fellow, considering that he is from Chicago.*

Jacobs had come to find customers for his "new-fangled light," and the man he found was Henry J. Rogers. Like most of his capitalist contemporaries, Rogers was a man of diverse interests. A native New Yorker, he had moved to Colorado as a young man, founding one bank there and another in Wyoming. He came to Appleton in 1873 as business manager for a group of Illinois investors. His chief responsibility was the Appleton Paper and Pulp

In 1882 the Rogers mansion in Appleton became "the world's first home lighted by hydroelectricity."
(Friends of Hearthstone, Inc.)

Henry Rogers, the industrialist who wanted the Edison light for his paper mill as well as his home

Company, a manufacturer of printing paper, but Rogers was soon involved in a canal company, a savings and loan, the local gaslight company, a pulp mill, and city politics as well.

Henry Rogers was building a house in 1882. Perched on the bluff overlooking his principal mill, it was taking shape as a Victorian showplace, complete with parquet floors, nine ornate fireplaces, and a rich variety of stained-glass windows. When Jacobs called on Rogers to explain the wonders of incandescent light, he found a ready listener. The river was already driving Rogers' mill. It could drive an Edison plant as well, providing light for his mill hands and the last word in home improvements for his mansion.

Rogers had never seen an Edison installation, but he purchased the Fox Valley rights to the system on the spot. He also persuaded three of his fellow capitalists to join him in the venture. Within days of Rogers' first meeting with Henry Jacobs, the Appleton investors received a visit from P. D. Johnston, Western Edison's chief engineer. Johnston outlined plans for a system that everyone found satisfactory, and the contract was signed on August 18. Work began almost immediately. The first dynamo was connected to a pulp beater in Rogers' mill. Wires were strung inside the mill, to a pulp mill downstream, and to Rogers' new home on the bluff. Local newspapers followed the work with intense interest. "Verily this is a progressive age," wrote the *Crescent*, "and Appleton keeps abreast of the times in everything tangible."

The Edison crew in Appleton was under strict orders to delay the plant's opening until their parent company's showpiece — the Pearl Street Station — made its debut. The Pearl Street switch was thrown on September 4, 1882. Twenty-six days later, on September 30, the lights flickered on in Appleton. The "system" was undeniably modest — two mills and a private residence — but the Appleton plant was the second Edison central station in the country and probably the first hydroelectric station anywhere. It also marked the beginning of generation in the Wisconsin Electric Power Company's present service area.

It quickly became apparent that one turbine could not drive both a pulp beater and an electric dynamo; the load was too variable to produce consistent lighting. Within a few weeks, a second dynamo was installed in a small shed near Vulcan Street, a half-mile downstream from the first. More customers signed on and, despite frequent short-circuits, the system was judged a success.

Henry Rogers was publicly enthusiastic about the system he had brought to life. When Edison publicists asked

The Vulcan Street plant that powered Appleton's "system" was little more than a shed.

The plant's operator kept a bucket of water handy in case the equipment overheated.

him to comment on the lights in his home, Rogers sent this testimonial:

> *I have about 50 lamps and have used them about 60 days. I am pleased with them beyond expression, and do not see how they can be improved upon. No heat, no smoke, no vitiated air, and the light steady and pleasant in every way, and more economical than gas, and quite as reliable.*

Privately, Rogers found the financial burden of his new home a source of worry. The lights alone cost him $60 every month, based on a charge of $1.20 per lamp. On April 10, 1884, Rogers wrote a somewhat plaintive letter to his employers in Illinois:

> *You will notice my account is overdrawn as it most always is. The reason for it this time is paying my taxes and life insurance. It is almost impossible for me to live on my salary since I built this fool house and I am pinching everything down to the last cent.*

Henry Rogers apparently found the money to pay his bills, and he remained an active figure in the electrical field. In about 1891, Rogers was among the incorporators of a firm that made incandescent lamps; an Edison customer became an Edison competitor. The Rogers family left Appleton in 1893, and their "fool house," a prominent local landmark by that time, passed through a succession of hands. In 1986, finally, the "Hearthstone" mansion was purchased by a nonprofit group committed to restoring the home and sharing its architectural treasures with the public. The Rogers residence, its original Edison fixtures still intact, is promoted as "the world's first home lighted by hydroelectricity."

In 1882 Milwaukee's Pabst brewery had 700 incandescent lights, half the total number in the state.

With the Appleton system up and running, Henry Jacobs was ready to tackle a significantly larger market: Milwaukee. When he arrived in the last months of 1882, Jacobs found a community of 125,000 people, by far the largest city in the state and the nineteenth-largest in the country. In the years following the Civil War, Milwaukee had evolved from a regional center of commerce to a national center of industry, and its rivers and railroad corridors were lined with tanneries, packing plants, flour mills, iron and steel plants and, of course, breweries.

Jacobs' first Milwaukee customer was, fittingly, a brewer: Capt. Frederick Pabst, owner of the largest brewery in a city already known for beer. A former Great Lakes skipper, Capt. Pabst was rebuilding his complex after a disastrous 1879 fire and, like Henry Rogers, he wanted the last word in lighting. Pabst contracted with Western Edison for a system of 700 lights, powered by underground wires from a dynamo on Juneau Avenue. Henry Jacobs supervised the installation, staying in Milwaukee for weeks and eventually making the city his home.

Always eager for publicity, Jacobs invited reporters to visit the work in progress at the brewery, and he took the opportunity to lambaste the arc light he had sold diligently for two years:

> *There have been at least twenty men killed by wires from the arc system, but we have yet to hear of the first accident from the incandescent system. Let them [arc companies] put their wires where they will be safe, and they cannot be safely placed over the ground.*

Addressing the "rumor" that underground wires leaked current, Jacobs, a

Patrons of Schlitz Park, a popular beer garden, could dance under arc lights that were visible 14 miles out on Lake Michigan.

masterful salesman with an arsenal of facts and figures, countered that the Pearl Street Station had 14,311 lights in service and lost only "48 millionths of 1 per cent" of its current in transmission. He also declared that Edison had the light of the future. Already, he said, there were 30,000 Edison lamps in use outside New York City, including 1,400 in Wisconsin.

The Pabst installation, with 700 lights, accounted for half the Edison lamps in the state. It dwarfed the central station in Appleton, and it was Jacobs' largest sale to date by a wide margin. It was not, however, the first incandescent system in Milwaukee. That distinction belonged to John Hinkel's saloon on Third and State, where Thomas Edison's marvelous lights came on in June of 1882.

The city already had an abundance of private arc light systems, some supplied by Brush Electric, Jacobs' old employer, and others by Thomson-Houston, the Cleveland firm that was Brush's chief competitor. The glaring lights were used to illuminate Milwaukee's leading hotels, opera houses, restaurants, railroad shops, and mills. One of the largest systems, installed in May, 1882, turned night to day at Schlitz Park, a popular Milwaukee beer garden. The Schlitz lights were visible 14 miles away, and lake ships began to use them as aids to navigation.

What Milwaukee lacked, surprisingly, was central station service. La Crosse, Fond du Lac, Appleton, Racine, and other Wisconsin communities had all entered the fray by 1883, but the state's largest city was, relatively speaking, in the dark. Henry Jacobs decided to change all that, and he used the private installation in Capt. Pabst's brewery as his springboard. In January, 1883, Jacobs invited Milwaukee's entire Common Council to view the brand-new Pabst system, offering to "furnish the conveyances for all who wish to attend." The invitation was promptly accepted, and the aldermen were impressed enough to award Western Edison (and its Milwaukee investors) a lighting franchise on February 5, 1883. Jacobs outlined a plan for three stations, one on each side of town, to avoid the expense of laying submarine cables in the city's rivers.

Any elation Henry Jacobs felt was short-lived. The Common Council had second thoughts after awarding the franchise, and the final document contained so many provisions and conditions that Western Edison, and Henry Jacobs, decided to drop the matter entirely. The obstructions continued a pattern: The city's aldermen had flatly rejected no fewer than six franchise applications since 1881, and they smothered two more attempts in layers of red tape. Lights were coming on all over Wisconsin, but three more years would pass before Milwaukee had its first central station.

Why the delay? The *Milwaukee Sentinel* (Jan. 13, 1884) opined that Milwaukee was "behind most cities in the matter of electric lighting" simply because "the citizens don't want the light yet." The paper continued: "There is nothing to regret in the situation. We are in the infancy of electric lighting and there is reason to expect great changes and improvements. When we do get a system it will be the best."

True enough, but other forces, notably political forces, were undoubtedly at work. The *Sentinel* was an unabashedly Republican newspaper. During an 1885 franchise debate, the *Milwaukee Journal,* representing the Democratic point of view, warned, "Watch out for the *Sentinel.*" The paper, they charged, was owned by the same people who owned the Milwaukee Gas Light Company, and it had become "the organ of the monopolists." The gas company was indeed a dominant Milwaukee institution. Since its formation in 1852, the firm had built a system totaling 90 miles of gas mains and 2,100 streetlamps. It provided light to thousands of private customers, and it had an exclusive contract to light the city's streets. The company's leaders had every reason to fear the coming of electricity. Their spokesmen appeared at every franchise hearing, arguing that electric light systems, particularly those relying on overhead wires, were pernicious threats to public safety. (Underground wiring, they knew, was prohibitively expensive for arc light firms.) The gas company lowered its rates in an effort to forestall competition, but the continuing parade of electrical applicants kept its lobbyists busy.

Those lobbyists found the Common Council an easy target. Like their counterparts in other large cities, Milwaukee's aldermen were unpaid, part-time officials with enormous power to award lucrative contracts. In the ethical climate of the Gilded Age, it was common for aldermen to have financial interests in, or even to work for, companies that did business with the city. Despite the occasional outcries of civic reformers, public officials were, as a group, subject to manipulation.

The power of the gas company, financial and otherwise, may well have led the Council to reject most electrical franchise applicants and to hamstring the others. Clearly, however, Milwaukee Gas Light's campaign was a holding action. After hearing the firm's usual plea for "safe" underground wiring during a franchise debate, one alderman prophesied, "Electric light is the light of the future, and even if it is buried under the ground, it will resurrect itself."

Badger Illuminating

One enterprise finally broke through the iron ring — the Badger Illuminating Company. Its name had nothing to do with its presence in the Badger State; the firm was named for its founder, Sheridan S. Badger, a banker's son who had set up shop as an "electric light broker" in Chicago. Working out of an office in the Loop, he organized lighting companies in Chicago, Joliet, and Racine, as well as in Keokuk, Iowa. In about 1884 Badger decided to enter the Milwaukee market.

His point of entry was the Industrial Exposition Building. Designed by Edward Townsend Mix, Milwaukee's leading society architect, the building was a Victorian fantasy, a glass palace three stories high and wider than a football field. Located at Sixth and Kilbourn (the present site of the Milwaukee Auditorium), it was built in 1881 as a showcase for local manufac-

Milwaukee's gas works in 1885. The gas company's lobbyists spent much of their time keeping applicants for electrical franchises at bay.

turers and their products. The industrialists held a sort of all-city trade show every autumn, with product displays and, in some cases, "model plants in actual operation." The building quickly became Milwaukee's first convention center. During the non-Exposition season, it hosted dairy fairs, German *Saengerfests* (singing festivals), and card tournaments, as well as concerts, horse shows, and charity balls.

S.S. Badger won the lighting contract for the Exposition Building in 1884, and his crews quickly installed a complete Thomson-Houston arc system. The installation satisfied everyone but Badger. The annual Exposition lasted only four or five weeks, and gatherings during the rest of the year were sporadic at best. The Chicago capitalist had a great deal of expensive equipment lying idle night after night during slow periods. The obvious solution was to turn the plant into a central station, supplying power to anyone who wanted it.The crucial question was just as obvious: How, given the Common Council's demonstrated reluctance, could Badger hope to secure the necessary franchise?

John A. Hinsey provided the answer. If other politicians had kept the city safe for the gas company, Hinsey was the politician who could make it safe for S.S. Badger. All but forgotten today, he was Milwaukee's undisputed boss in the 1880s, a scalawag in an age of scalawags. Hinsey's role in the electrification of the city, for lighting and more particularly for traction, was seldom salutary, but it was absolutely pivotal.

John Hinsey came up the hard way. He was born in 1833 in Lebanon, Pennsylvania, one of eight children.

The Industrial Exposition Building, a Victorian showplace that doubled as Milwaukee's first central electric station (Milwaukee Public Library)

His father died when he was eight, and the boy was "bound out" to a Quaker farmer until the age of 18. "Young Hinsey," wrote a biographer, "became early accustomed to the idea that effort must be the necessity of his existence." As a young adult, he moved to Indiana and then, in 1865, to Milwaukee, where he landed a job with the Chicago, Milwaukee & St. Paul Railroad (later the Milwaukee Road). Hinsey became the railroad's chief claims agent, but he enjoyed far more notoriety as a politician. His career began in 1871, when he was elected to the city council from the South Side. Within five years, Hinsey had reorganized the local Democratic Party and turned it into a machine whose levers he controlled. He was known to friend and foe alike as "Boss" Hinsey, keeping his minions in line with patronage, favors, and simple persuasion. He moved to a well-heeled section of the East Side in 1879 and easily won election as its alderman. In 1885, not long before Badger applied for a fanchise, Hinsey was elected president of the Common Council.

Boss Hinsey became Sheridan Badger's champion – for a consideration. (Altruism had never been one of Hinsey's more conspicuous traits. He was once charged with embezzling funds from his own Knights of Pythias lodge.) Although the terms of his deal with Badger were never disclosed, the boss was soon identified as a part-owner of the lighting company. Hinsey railroaded the Badger franchise through the Common Council, carrying it to approval on November 21, 1885, by a vote of 26 to 6. The ordinance authorized overhead wires, prohibited Milwaukee Gas Light from buying an interest in the firm, and made provision for municipal purchase of the Badger plant at some future date. The issue of municipal ownership surfaced early.

The Republican *Sentinel* howled when the Badger ordinance was passed. "A Nondescript Company Given a Valuable Franchise," shouted the headline. The paper dismissed S.S. Badger as "an unknown and irresponsible party" and castigated Hinsey as "an irresponsible and incompetent political boss" who "had his supporters well in hand." The Democratic *Journal* was more sympathetic; "Let There Be Light," proclaimed its headline.

How decisive was Boss Hinsey's role in the Badger debate? Milwaukee Gas Light's plant was located in the Third Ward, a heavily Irish neighborhood south of downtown. Many Third Warders worked in the gashouse, and they viewed any threat to their employer as a threat to their own welfare. When the Badger franchise was approved, it was John Hinsey, not Sheridan Badger, they hanged in effigy.

With the coveted franchise in hand, Badger and his associates went to work. One of his spokesmen promised "everything the public wants, from the 2,000-candlepower arc lamps for lighting the streets ... to the one-half candlepower incandescent lamps used by doctors in examining a patient's throat." The company never delivered incandescent lights of any kind, but Badger Illuminating did begin to build an arc system in the early weeks of 1886. The first step was to string wires from the Exposition Building to customers nearby: the German shop keepers on Third Street, the Turner Hall on Fourth, and other establishments in need of large-area lighting. The Badger crews encountered obstacles almost immediately. They took the most direct routes to their customers, often attaching wires to private homes and businesses along the way — without asking permission. When one property-owner complained, he was told that "Hinsey authorized it." Some residents simply stepped outside and cut the wires down. After several days of friction (and hundreds of yards of scrapped wire), the crews found alternate routes, and the wire-stringing continued.

"Boss" John Hinsey, the scalawag who routinely used his political power to further his business interests

Badger Illuminating expanded rapidly, and within months the system outgrew its quarters in the Exposition Building. In July, 1886, the company moved into a rebuilt flour mill on the Milwaukee River, at the foot of McKinley Avenue. Steam engines were installed in the basement, power trains on the first floor, and dynamos on the second. The move was, in retrospect, a wise one. The Exposition Building burned to the ground in 1905, but the Commerce Street Power Plant (which followed Badger on the riverfront) remained in service until 1988 — a regional record for continuous service on one site.

With a new central station and an aggressive marketing campaign, Badger Illuminating quickly expanded its service area. Overhead wires remained in place on the West Side, but submarine cables and concrete conduits carried power to the East and South Sides. The system received an important boost on December 15, 1887, when the Common Council awarded Badger a three-year contract to light the city's streets. By the middle of 1889, the company had erected 185 streetlights; by the end of 1890, there were 402.

Sheridan S. Badger followed the progress of his Milwaukee enterprise from an office in the Rookery, the celebrated new "skyscraper" in Chicago's Loop. Edward C. Wall was the president of Badger Illuminating of Milwaukee, and his presence underscored the importance of political connections. Like John Hinsey, Wall was a full-time party boss, and he exercised nearly as much control over the state Democratic Party as Hinsey did over its local branch. Day-to-day management was the responsibility of Harry Boggis. He supervised a staff of perhaps 15 people, from office workers to electricians. New employees generally started at the top, climbing dozens of towers each day to trim the arc lamps.

By the late 1880s, the Milwaukee skyline was studded with light towers. "Electricity is becoming general for street illumination," the *Sentinel* could report in 1888, although the number of lights would more than double in the next two years. Arc lights, continued the paper, helped to guide "the unsteady steps of the beer-hall loiterer" at midnight, and they showed "the measure of the enterprise of the community." Only one group was ill-served by the lights: servant girls, "whose good-night squeeze at the gate has all the romance knocked out of it in the white light of the electric lamp."

Breaking the Monopoly

Badger Illuminating enjoyed a virtual monopoly in Milwaukee for more than three years. The city's growth had continued in the meantime, and so had its appeal as a market for electrical manufacturers. The parade of electrical franchise applicants continued, and the wisdom of an exclusive contract was increasingly open to question. Even someone as well-connected as S.S. Badger could not hope to maintain a monopoly forever.

A tragedy helped to loosen Badger's hold. In the first week of May, 1886, thousands of Milwaukee workers went on strike, demanding an eight-hour day without a cut in pay. (The norm at the time was 10 to 12 hours, six days a week.) They marched through the city, closing scores of factories, but one plant remained open — a huge iron and steel mill in Bay View, an industrial suburb just south of the city limits. On the morning of May 5, nearly 1,500 workers marched on the mill. The state militia was waiting. When the crowd ignored (or failed to hear) the militia captain's command to halt, he ordered his troops to open fire. At least seven people were killed.

The Bay View incident was the bloodiest labor incident in the state's history, and it galvanized Milwaukee's blue-collar voters. The pro-labor People's Party was organized soon after the shootings, and laborites took 10 Common Council seats in the 1888 elections (against 14 for the Republicans and 12 for the Democrats). A reform movement was beginning to take shape. Boss Hinsey remained a potent figure, beginning his second term as Common Council president in 1888, but the city's aldermen were developing minds of their own.

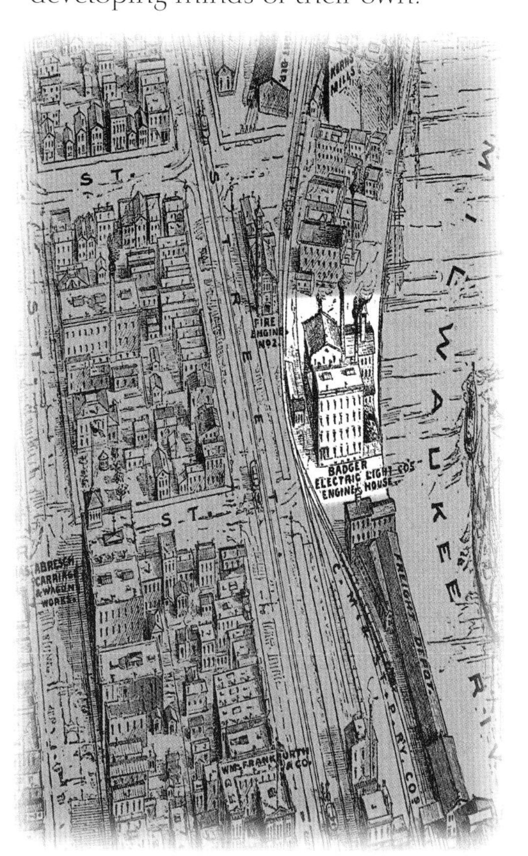

Badger Illuminating's powerhouse was a converted flour mill on the Milwaukee River.

As the political climate changed, the pressure for competitive franchises mounted. New ideas received a wider hearing, and a familiar figure presented one of the most novel. In June, 1888, Henry Jacobs, the peripatetic Edison agent and by now a city resident, appeared before the Common Council. Why, he asked, should the city award lucrative franchises to private companies when it could operate its own light plant? He presented a detailed plan for a central station (Edison, of course) that Milwaukee could run for the benefit of its citizens. The plant, Jacobs estimated, would cost $250,000, and operating expenses would total $65,000 a year — far less than the $312,000 needed for the same illumination from gaslights. The salesman argued that the "special advantages" of the incandescent system (among them "instantaneous lighting and extinguishment") would enable the city to pay for the plant within three years.

A municipal ownership faction formed quickly, led by reform alderman Henry Baumgaertner. In November a Common Council committee asked the Wisconsin legislature to authorize a bond issue for the plant. After a brief period of study and debate, the matter was quietly dropped. The issue was dead, for the time being, but one of Baumgaertner's arguments demonstrated how essential electric lighting had become in just a few years. "If the city furnishes water to its inhabitants," said the alderman, "it ought also to furnish light without having to purchase it from an outside corporation."

Henry Jacobs' pitch to the Common Council was one of his last public appearances. He continued to represent Edison in Wisconsin, but Jacobs, like many another promoter, had gone into business for himself. He was a principal in several Milwaukee companies, including one that sold electric thermostats and another that manufactured electric fire alarms. By 1890, however, Jacobs began to lose his long-term battle with diabetes. In the last weeks of the year, he moved to his father-in-law's Fond du Lac home for treatment and rest, but his health continued to fail. The godfather of Wisconsin's electrical service died on April 7, 1891. He was 36.

The Edison interests continued to seek a foothold in Milwaukee. When it became apparent that the city was not about to buy a lighting plant, they decided to apply for a franchise of their own. The Edison Electric Illuminating Company was organized in July, 1889, succeeding the stillborn affiliate Henry Jacobs had formed in 1883. Edison General Electric, the parent company, held its customary 30-percent share of the local company's stock, and Edison GE dispatched an executive from New York to drum up support. (He also offered to deposit $500,000 in any local bank.) Brewer Fred Pabst, banker Charles Ilsley, lawyer John Van Dyke, and other local luminaries signed on, adding prestige (and capital) to the enterprise. The new company applied for a franchise almost as soon as its incorporation papers were filed.

Predictably, John Hinsey and S.S. Badger led the opposition. Amid allegations that Badger "owned" some aldermen, Hinsey denounced the Edison affiliate as "a straw company" and the Edison system as "good for nothing." Badger himself made an appearance. "I know that the Edison system is no good," he said, "for it proved a dire failure in Chicago." (That news might have come as a surprise to the founders of Commonwealth Edison.) The issue came to a vote on September 23, 1889. Boss Hinsey fulminated for three hours, giving what the *Milwaukee Sentinel* called "the speech of his life." It was to no avail. By a vote of 28 to 6, the rebellious Common Council authorized Edison Electric Illuminating to offer central-station service on the same terms as the Badger company. The *Sentinel* was exultant. "Boss No Longer ... Hinsey Knocked Out in the Council," read the banner headline. The *Milwaukee Journal* charged, rather quietly, that Hinsey's opposition "has, by the exercise of considerable cunning, been made to appear as designed to shut out competition."

In the summer of 1890, Edison Electric began to build an imposing power plant on the Milwaukee River, just north of Oneida (now East Wells) Street. The county poorhouse and the municipal morgue were next door, and City Hall was only a block away. Construction proceeded rapidly, and the plant began to offer limited service as early as October. Within two years, it was generating power for 10,000 incandescent lights.

Granting the Edison franchise opened the floodgates. After the political upheaval of 1888, franchises were given to practically every company that asked, and the city's street-lighting business was divided among them. Milwaukee had three central stations

In the summer of 1890, Edison Electric built a state-of-the-art power plant on the east bank of the Milwaukee River.

in 1890 — Badger, Edison, and Milwaukee Electric — and more were on the way. The unforeseen result was absolute chaos. Different companies erected their poles on the same streets, and some served different customers in the same buildings. The wires, and their owners, were soon hopelessly tangled, and tensions frequently rose to the boiling point.

Powerful forces of convergence were already at work. Unknown to most of the local principals, an eastern tycoon and a Wisconsin political boss were working quietly behind the scenes to bring order out of chaos. They envisioned a unified electrical system serving the entire Milwaukee region, and the focal point of their planned empire was the Edison plant on the Milwaukee River. That plant was nominally built to power a lighting system, but lighting was only one side of the system the entrepreneurs hoped to forge. Electrified transit was the other. It had taken years for the lights to come on in Milwaukee, and the process was marked by delays, intrigue, and endless controversy. The other side of the story — traction — is every bit as complex and, if anything, even more colorful.

II. Farewell to Dobbin

Until the late 1800s, urban transit meant horse-drawn transit. The first "horse railroads" appeared on the East Coast in the 1820s, and the innovation gradually spread west, reaching Wisconsin on the eve of the Civil War. In 1859 George Walker, one of Milwaukee's three founding fathers, organized the River and Lakeshore City Railway Company. Milwaukee had fewer than 50,000 residents at the time, but Walker was confident that a horsecar line serving the affluent East Side would attract ample patronage. Omnibus companies, who had enjoyed a monopoly on passenger service for more than a decade, objected strenuously, but Walker's enterprise appealed to local pride. "The city that is without its street railroads now-a-days," wrote the *Milwaukee Sentinel,* "is regarded by travelers as being decidedly behind the age." City officials granted Walker a franchise, and his first cars rumbled down Water Street on May 30, 1860. The River and Lakeshore line thus became the earliest corporate predecessor of the Wisconsin Electric Power Company.

George Walker's horsecars attracted crowds at first but, as the novelty wore off, so did the firm's financial luster. After years of losses, the line was reorganized as the Milwaukee City Railway in 1865. The losses continued until 1869, when Isaac Ellsworth bought the line and finally made it a paying enterprise. A native of New York State and a carriage-maker by trade, Ellsworth proved himself an aggressive manager — so aggressive, in fact, that he once got into a fistfight with an ice-wagon driver who refused to push his disabled rig off the streetcar tracks. The new owner's first step was to tear up the underutilized tracks on the East Side and put them into service on the west side of the river. Ridership increased rapidly, and Ellsworth soon extended his line to the South Side.

Encouraged by the Milwaukee City Railway's success, other entrepreneurs tried to emulate Ellsworth's operation.

With ample backing from local capitalists, it was ready for service by fall.

The Cream City Railroad, organized in 1874, restored horsecar service to the East Side and soon added a line to suburban Bay View. The Cream City line's president was Winfield Smith. A native of Green Bay, Smith was a prominent Milwaukee lawyer who had served four years as Wisconsin's attorney general. Smith was an ardent Republican, but he had the political services of Edward C. Wall, the Democratic boss who would later become president of Badger Illuminating.

Milwaukee's third horsecar line, the West Side Street Railway, began operations in 1875, serving the mix of wealthy and working-class communities west of Milwaukee's downtown. Its guiding light was Washington Becker. Born in New York State and educated at Harvard, Becker came to the city as an aspiring young lawyer, but urban transit soon absorbed most of his time. After acquiring a controlling interest in 1880, Becker devoted his full attention to the West Side line, and he earned a reputation as the most astute transit manager in the city.

By 1880 Milwaukee boasted three horsecar companies operating on nearly 30 miles of track. Although the lines practically converged at the heart of the city, each had its own territory. Milwaukee City, easily the largest line, served the North and South Sides; Cream City controlled the lakeshore districts from the East Side to Bay View; and West Side, of course, had a monopoly on West Side traffic. There were no attempts to integrate the systems; passengers traveling from the East Side to the West Side paid a nickel every time they changed cars. Frugal crosstown commuters learned to walk.

Although they became institutions, the horsecar lines were less than adequate to the needs of a growing city. In the two decades following the debut of George Walker's line in 1860, Milwaukee's population soared from 45,246 to 115,587 — an increase of 155 percent. Thirty minutes was considered the maximum travel time to downtown, and the speed of the horsecars (six miles per hour) therefore limited the city's expansion to a radius of three miles. Civic leaders and newspaper editors, not to mention real estate developers, began to agitate for a new form of rapid transit."

Lack of speed was not the only drawback of horse-drawn transportation. Equine epidemics ("epizootics," as they were called) sometimes wiped out entire stables, bringing streetcar service to an abrupt halt. But the horsecar system's most visible drawback was undoubtedly manure — mountains of it. In 1881 the three Milwaukee companies had 640 horses and mules in service. Each produced 10 to 20 pounds of manure each day

A Milwaukee City horsecar on the Third Street line in 1882

George Walker, a Milwaukee pioneer who launched the city's first horsecar line in 1859

(Milwaukee Public Library)

— a total of up to six tons. No one considered diapers, and very few farmers took advantage of the bountiful supply. Rains simply carried the waste to the nearest river, contributing to a pollution problem of staggering proportions. An 1881 visitor called the Milwaukee River "a currentless and yellowish murky stream, with water like oil, and an odor combined of the effluvia of a hundred sewers." The problem was most acute in the vicinity of the car barns. In 1889 the mound of manure behind the Cream City barns on Kinnickinnic Avenue caught fire and smoldered for days. "The manure is burning and emitting a malodorous smoke," reported the *Sentinel*, "endangering the safety as well as the health of the neighborhood."

The time for a new technology had come, but electricity was not the first choice. In 1877 several Cream City Railroad officials opened a summer line to Forest Home Cemetery, a popular South Side recreational spot in the days before public parks. The line's motive power was a small steam engine, which a reporter described as "almost noiseless ... and more readily managed than a team of horses or mules." A stuffed mule was mounted on the front of the engine to give passing horses a sense of security. The "steam dummy" was a novel addition to the city's transit system, but it proved more costly than animal power. The Cream City line replaced its stuffed mule with live teams in 1879.

Charles Van Depoele introduced electrified transit as a more permanent solution. Born in Belgium and based in Detroit, Van Depoele was Frank Sprague's chief competitor in the early years of electrified transit, and he was the first of the pair to reach Wisconsin. In 1886 Van Depoele answered a call to Appleton. The Fox River city had already established itself as a national trailblazer in hydroelectric power, and its leaders showed just as much foresight in the field of traction. On January 19, 1886, a group led by Judge Joseph Harriman incorporated the Appleton Electric Street Railway. Construction began in spring, the first Van Depoele-powered car arrived on August 12, and regular service commenced four days later. (The spellbound spectators included a visiting circus troupe.) "Curves and grades are overcome without any difficulty," reported the *Milwaukee Sentinel*, "and the success of the enterprise seems to be demonstrated." The Appleton system was the first electrified transit operation in the state and one of the first in the nation.

Transit pioneers: Van Depoele electric streetcars whiz down College Avenue in Appleton.

Following in Henry Jacobs' footsteps, Charles Van Depoele used his apparent success in Appleton as a springboard to larger markets, including Milwaukee. In the middle of August, 1887, he built a two-block demonstration line in Milwaukee, choosing a steep section of Seventh Street between State and Juneau. Van Depoele put up poles and wires, connected his dynamo to the steam engine in a nearby stove foundry, and was soon ready to show off his invention.

The first run provided a good deal of excitement for a crowd estimated at several thousand. One of the overhead trolleys fell to the tracks during the demonstration. "The result," reported the *Sentinel*, "was a series of reports, as if a bunch of firecrackers had been exploded, causing a stampede among the spectators." But there were more surprises in store. Stephen Bell, the West Side Railway driver chosen to pilot the new car, described his maiden voyage to a reporter:

> *The car struggled up the hill, but at the top the brake failed and it started to roll back down. There was a load of people on the car. They got pretty excited. But I kept the brake on and the car stopped about halfway down. Then I turned the power on and up we went again. This time it worked.*

Bell left the streetcar business shortly after the test run, retiring to a more sedate life on his family's Saukville farm.

Charles Van Depoele's intent, of course, was to interest Milwaukee's three horsecar companies in converting to his system. Electric cars, the inventor said, cost only pennies a day to operate, while horsecars cost six to seven dollars, and he mentioned three-cent fares as a distinct possibility. Van Depoele found no takers. Political maneuvering had delayed the opening of the city's first electric lighting plant for years, and electric transit would suffer the same delays. Milwaukee, in fact, was in the first stages of a traction war in 1887, a war that would rage for more than three years. It was universally agreed that some other motive power — steam, electricity, cable, or even compressed air — would soon replace horse-drawn streetcars, but there was spirited disagreement about which technology would prevail, and who would be allowed to use it first.

A promoter with the unlikely name of Leander F. Frisby fired the first shot in the traction war. Frisby was a prominent Republican serving his fifth year as the state's attorney general, but he also had extensive business interests. He owned the rights to a cable car system and, on April 19, 1887, Frisby applied for a franchise to put that system to work in Milwaukee. Cable cars were a national rage at the time. Usually associated with the hills of San Francisco (where they first appeared in 1872), the cars had turned up in numerous other communities. Chicago, a notably flat city, introduced a cable transit system in 1882, and it quickly grew to 500 cars running on 86 miles of track. The principle was the same in every installation. The cars were attached to an endless cable buried in a casing between the tracks. When the "grip man" wanted to stop for passengers, he simply released the cable clamp until he was ready to proceed. A central steam engine kept the cable moving, and a system of underground pulleys enabled the cars to make even the sharpest turns. Construction costs were extremely high, and the fate of every system hung by a single steel thread, but cable cars were a marked improvement over horses and mules.

Frisby and his partners sought approval for a cable line extending from Lake Michigan to Wauwatosa, a suburban settlement then two miles west of the city limits. The route cut through the territories of all three horsecar companies. They had done

The Milwaukee City line's carbarn stood on Second Street just north of Grand (now Wisconsin) Avenue.

little to update their systems, but the existing companies were outraged by this new threat to their livelihoods. They quickly sought additions to their own franchises, and their lawyers condemned cable systems as a threat to public safety. The Chicago system, charged a horsecar lobbyist, killed at least one citizen every day.

The older transit companies gave Frisby a cool reception, but his real enemy was another cable firm, one established only weeks after his. The Milwaukee Cable Railway Company was the creation of Francis Hinckley, a Chicago capitalist who controlled a modest railroad network in northern and central Illinois. The company's resident manager was none other than "Boss" John A. Hinsey, the head of Milwaukee's Democratic machine. (Hinckley reportedly gave Hinsey 40 percent of Milwaukee Cable's stock in return for his services.) Hinsey had engineered the Badger Illuminating franchise in 1885, and he was determined to do the same for the Milwaukee Cable Railway.

The summer of 1887 was a watershed in Milwaukee's transit history. The Common Council considered franchise requests from two rival cable companies. Charles Van Depoele showed off his new electric car. A short-lived electric railway company was incorporated. Construction of a steam-powered line to suburban Whitefish Bay began. Besieged on every front, the horsecar companies fought back vigorously, and the result was a free-for-all. The *Milwaukee Sentinel* (Aug. 14, 1887) described the situation as a "street car epidemic." The paper continued:

Dressed for any weather, Milwaukee City drivers gather with whips in hand.

> *There is no street so obscure, no grade so steep, that the "enterprising capitalist" will not accept for a cable or horse car line. It has become the set policy of the companies to take all they can get, if for no other object than to prevent others from getting it.*

The summer's central contest was the cable franchise fight between the Frisby interests and the Hinsey-Hinckley partnership. Debate in the Common Council was, to say the least, spirited. "Order Ran Riot," screamed the *Sentinel* headline for August 30, describing "Beer Garden Scenes In The Common Council." "Howling aldermen" impugned each other's honor, leading the paper to remark that they acted "entirely as though they were the paid agents of different companies." That was undoubtedly the case. Boss Hinsey's championship of his own company was, of course, an outrageous conflict of interest, but it was hardly exceptional. In the heat of the franchise debate, Alderman Garrett Dunck, who had close ties to the city's horsecar interests, was accused of "boodling" (accepting bribes) by one of his colleagues. Dunck's angry reply spoke volumes about the ethics of the Gilded Age: "Alderman Richards is of so little importance that he is not worth a cent to anybody to buy him." In the Milwaukee of 1887, that might have been the worst thing one politician could say about another.

Given Hinsey's clout (and Hinckley's purse), the outcome of the contest was hardly surprising. The Milwaukee Cable Railway's franchise was approved in October, and Frisby went spinning off to other ventures. The "Hinsey line," as it was soon called, was authorized to run cable cars from the east side of downtown to the western city limits. The horsecar companies had managed to insert some restrictions in the franchise, including a time limit on construction, but Hinsey's way was clear. "The city needs rapid transit," declared the *Sentinel* (Aug. 31, 1887), and John A. Hinsey emerged as the man most likely to lead Milwaukee into a new era.

Hinsey's grand entrance dissolved into a comic opera. He had sworn that his crews would complete the new line in 90 days, but the Milwaukee Cable firm did not lay a single section of track through the remainder of 1887 and all of 1888. When the deadline for completion passed, Hinsey wangled an extension. By 1889, however, he had abandoned the cable system in favor of electric cars powered by storage batteries. All along the proposed route, reported the *Sentinel* (July 3, 1889), "the only topic of conversation is the promised arrival of the electric motor car." Weeks passed with no cars and no visible progress of any kind, but Hinsey was the picture of confidence. "I'm going to have nice big cars," he promised in July, "with wide aisles and a broad platform for smokers. The conductors will all wear uniforms and Milwaukee will have a high-class street railroad."

But the politician soon lost interest in the storage battery system for his "nice big cars." In August he sent a group of aldermen to Boston and other eastern cities, ensconcing them in a private Pullman provided by his employer, the Milwaukee Road. The trip, said Hinsey, would give city officials "a chance to try everything in the electric motor line." He promised to follow their recommendations, but Hinsey announced a new plan days before the group returned. The new line, he declared, would use the overhead electric system already at work in cities from Bangor to Seattle. The Milwaukee Cable Railway Company, accordingly, was rechristened the Milwaukee Electric Railway. The exasperated *Sentinel* (Aug. 8, 1889) predicted that Hinsey would soon ask for a franchise using "hydraulic or balloon or some other power." "We imagine," added the editor, "that wind alone will be used for Mr. Hinsey's line."

Boss Hinsey's hold on the city's transit policy was remarkable. He was able to keep his competitors at bay for nearly two years without so much as one pole in the ground. The first construction deadline passed, and then the second, but Hinsey had little trouble securing extensions. In 1888 he sought permission to expand his phantom line to the South Side, the East Side, and Bay View. Horsecar companies were already serving those areas, but the Common Council gave Hinsey a blanket franchise covering 30 miles of streets. Winfield Smith, head of the Cream City line, described the Council as "surprisingly credulous and partial to the new Company," and that

The Milwaukee & Whitefish Bay Railway, powered by steam, did a brisk resort business on summer weekends.

was an understatement.

Soon after the boss began his second term as Council president in 1888, the older companies began to complain of "harassment" by the city. The aldermen denied them line extensions, raised their license fees, and required longer hours of operation. The new rules prompted one frustrated horsecar official to issue a classic statement of the Gilded Age business philosophy. "A street railroad," said Frank Bigelow of the West Side line, "is as private an institution as a dry goods store, the affairs of which are of no concern to the public." He was dead wrong, as later events would illustrate, but Hinsey's machinations sparked understandable resentment.

Hinsey's total control was all the more remarkable in light of a reform movement gaining strength in the Common Council. Laborites elected in the wake of the 1886 Bay View shootings had minds of their own, but this vocal minority generally let Hinsey have his way. Why? The most likely reason is that they wanted rapid transit. Throughout the franchise debates, one central argument recurred: Rapid transit would open the city's outskirts to the workingman. The *Sentinel* opined that faster streetcars would "place the poor man along the street railway on a par with the rich." The city attorney favored the Hinsey line as a means to give "our wage workers" access to "plenty of pure air and God's own sunshine thrown in." Hinsey himself stated that his line would enable ordinary workers to "reach the city limits and own houses of their own." When electric streetcars finally came, they tended to benefit the middle class more than wage-earners. Rapid transit, however, became a sort of Holy Grail for some reformers, even if it came at the hands of someone as distasteful as John A. Hinsey. "I am charged with being dilatory," he said, "but I will give the people as good a street railway as there is in America."

Milwaukee did not sit still while Boss Hinsey flounced from one plan to the next. More than 150 American cities had electric street railways by the end of 1889, and civic leaders were chagrined that Milwaukee was not among them. One after another, local entrepreneurs tried to break Hinsey's apparent hold on the new technology. In 1889 a group backed by the Pabst brewing family announced plans to build an electric line between Milwaukee and Wauwatosa. "It will not," said a spokesman, "be another case of Hinsey railroading. The gentlemen taking a hand in the matter are too well-known and reliable." The Milwaukee and Wauwatosa line finally went into operation in 1891.

The old horsecar companies had seen the handwriting on the wall several years earlier, and all were making their own plans to electrify. (One firm cited "cleanliness, comfort, and beauty of the streets" as reasons to convert — an oblique reference to the manure problem.) Washington Becker, head of the West Side Street Railway, was probably the most aggressive transit manager. After examining several systems, he settled on the Sprague-built streetcars that were becoming the national standard. All he needed was the city's permission to electrify.

Hinsey, meanwhile, was experiencing financial problems. He had wrested control of the Milwaukee Electric Railway Company from his Illinois partner — while he was Common Council president — but Hinsey lacked the capital to begin construction. When New York banks refused to loan him money, the boss turned to Charles Pfister. Known to nearly everyone as "Charley," Pfister was treasurer of the Pfister & Vogel tanning firm, principal owner of the Whitefish Bay steam railway, and one of the state's leading Republicans. He apparently found the Democratic boss's company congenial. In exchange for bonds that gave him effective control of the Milwaukee Electric Railway, Pfister forwarded the necessary funds to Hinsey.

With a new source of capital and growing pressure from his rivals, Hinsey finally began to build a street railway — nearly two years after receiving a franchise. His crews started to lay track in August of 1889, and Hinsey purchased an old flour mill on the Milwaukee River for use as his power plant. It adjoined the Badger Illuminating plant to the south. Badger was headed by Edward Wall, Hinsey's counterpart in the state Democratic Party, and Hinsey himself had a substantial interest in the firm. Separated from the beginning, electric lighting and electric transit were beginning to merge.

When it became clear that Hinsey was actually making headway, the *Milwaukee Sentinel* (Sept. 22, 1889) dropped, for at least one issue, its customary opposition to the boss:

> *To John A. Hinsey the eyes of the people of Milwaukee are now turned. From him they expect relief from the horses and mules, and of him they think when they speak of rapid transit. Sitting upon a hard chair in a dimly lighted room at the northwestern*

A transformation in transit: The West Side streetcar line, long Milwaukee's smallest …

end of the city, the laborer dreams of the mornings when he will be carried to work in Hinsey's warm and roomy motor cars with the speed of twenty miles per hour, and, reposing upon a soft sofa in a glittering parlor at the western limits of the city, the business man paints to himself the noon hours when Hinsey's elegant cars will take him home for dinner every day, without loss of time. …All expect from Hinsey their salvation.

But Boss Hinsey soon had company. The reform faction on the Common Council had been gaining strength and, in the autumn of 1889, a majority of aldermen finally broke ranks with their president. On September 23, the Council approved the Edison Electric Illuminating franchise, ending S.S. Badger's lighting monopoly. On October 14, with Hinsey's line still under construction, the aldermen authorized every horsecar company to convert to electricity. The race was on.

With his stranglehold on transit policy broken, Hinsey had little choice but to get down to business. In rushing his line to completion, however, the city's would-be streetcar czar showed genuine ineptitude. Hinsey used the cheapest materials and the least competent workers available. Poles were held in place with braces and guy wires, and a jerrybuilt section of his overhead system was felled by a load of hay. A *Sentinel* reporter (Jan. 17, 1890) observed that numerous poles on Hinsey's line were secured to nearby trees, "to the great detriment of the trees and little apparent benefit to the poles." City officials ordered the work redone, and Hinsey lost valuable time.

Washington Becker, by contrast, did it right the first time, anchoring the West Side Railway's poles in six feet of concrete. Becker and Hinsey were neck and neck for a time, but Becker, working through the winter months, was ready to unveil his new system by spring. Milwaukee's first electric streetcar, a Sprague model, made its triumphal maiden voyage down Wells Street on April 3, 1890. The West Side Street Railway, long the city's smallest, had won the electrification derby.

Every age has its wonders. To the Milwaukeeans of 1890, the streetcar was just as marvelous as manned space flight would be to a later generation. The *Sentinel* (April 4, 1890) described the first horseless run on the Becker line:

The application of the trolley to the wire drew from the latter a stream of bluish-red sparks and from the spectators a chorus of "oh's." Some of the people who had gathered to see the start were frightened and as cautious as an Indian who had never seen a locomotive.… It was a circus for the small boy and a spectacle of no little interest even

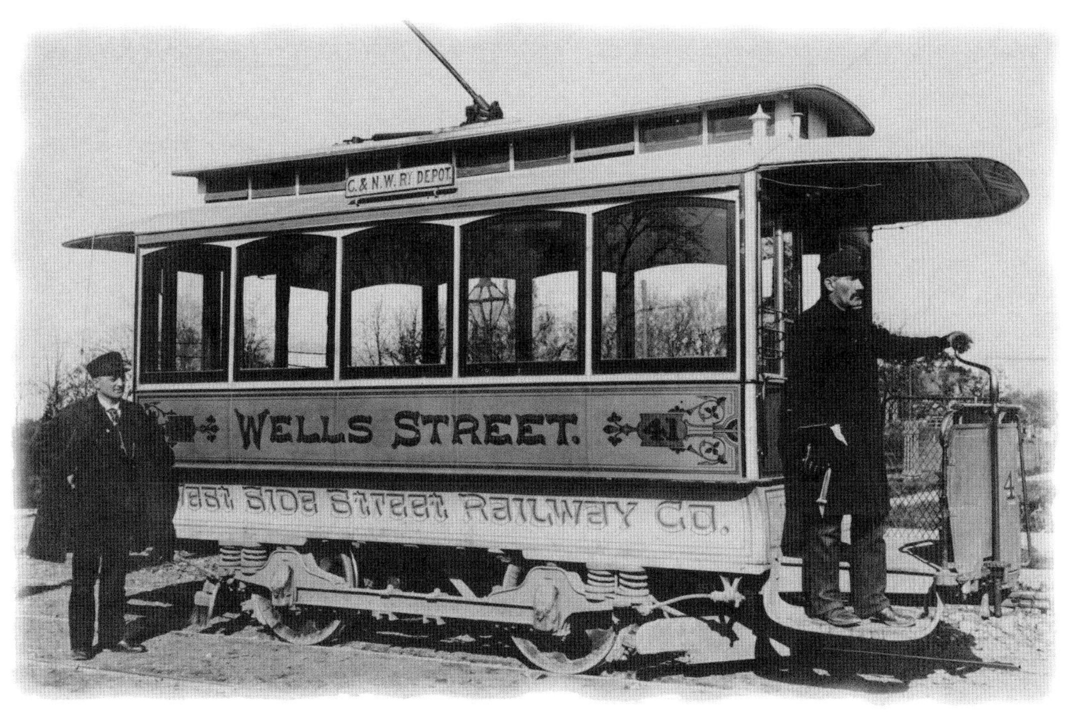

... won the race to electrify in 1890.

to those who have lived to see the telegraph and the telephone put into practical use.

Becker's cars were soon whizzing down the city's streets "like streaks of lightning on roller skates," but his chief rival was not far behind. Hinsey's line opened for business on April 20, roughly two weeks after Becker's. With a carnation in his lapel, the boss watched the proceedings from behind a team of matched grays. His system's debut was less than triumphant. Two particularly crooked poles fell into the street during the maiden run, causing a long delay. Most of the passengers were boisterous baseball fans heading to a game, and they rocked the cars so violently that the scrapers on both ends snapped off.

Despite the opening-day calamities, the "Hinsey line," serving neighborhoods north and west of downtown, soon became a well-used link in Milwaukee's transit network, and electrification of the remaining lines proceeded at a brisk pace. "Horses Sold Cheap," read a *Sentinel* headline as the work progressed. George Peck, a humorist-turned-politician, became the city's mayor in 1890. Shortly after his election, the author of *Peck's Bad Boy* made a light-hearted boast about his city's progress: "Ordinary places have nothing but old plug horses to pull their cars around, but with us, you can now ride in a car with nothing in front."

The same election that swept Peck into office saw the departure of Boss Hinsey. The Milwaukee Road had moved its headquarters to Chicago shortly before. Hinsey was apparently unable to live on the income from his various business interests, and his political power was eroding before his eyes. The politician elected to follow his employer south. Hinsey commuted from Milwaukee for a time, and he retained his interest in the Milwaukee Electric Railway, but his political career came to an abrupt end. John A. Hinsey was a boss no longer.

The post-Hinsey Common Council adopted a laissez-faire approach to electrical franchises, for both lighting and traction. Few applicants were turned away. Just as competing light companies served customers in the same building, rival streetcar lines had franchises for the same streets. They tore up each others' tracks and tore down each other's wires with regularity, and the *Sentinel* carried numerous stories about the "merry railway war." On one downtown street, the Milwaukee City Railway controlled the tracks and the Milwaukee and Whitefish Bay Railroad owned the wires, creating a stand-off that benefited no one, least of all the city's transit riders.

Chaos reigned supreme in 1890. Milwaukee's electrical firms were a bundle of loose wires, connected haphazardly and arcing dangerously. Both traction and lighting, the day and night loads of a true electric utility, were snarled in a jumble of competing interests. Help, however, was on the way. A capitalist and a politician were poised to clean up the mess, to transform the loose ends and short circuits into one coherent system controlled by one cohesive management. Their story, in its fundamental outline, is a tale of two Henrys.

Chapter 2

1890-1896

A Tale of Two Henrys

Through the years that John Hinsey sat on his franchise and rival businessmen gnashed their teeth, a subplot had been developing. It was clear that electricity was destined to be the motive force for both lighting and transportation. It was equally clear that efforts to electrify Milwaukee were pathetically disjointed, crippled by political shenanigans and competition between entrenched financial interests. The situation seemed to cry out for order, and it called just as loudly for capital. The seeds of a system were present in Milwaukee, but they would not sprout until someone showered them with money.

Henry Clay Payne saw the potential bounty in the city's unkempt electrical field. He had come to Milwaukee from Massachusetts in 1863, a 20-year-old Yankee with 50 dollars in his pocket. He showed little aptitude for business in his early years, trying and failing three times to establish dry goods stores, but the failed shopkeeper found singular success in his true calling: politics. In 1872 Payne organized the Young Men's Republican Club, a group that would breathe new life into a moribund party organization. A tireless worker in both local and national campaigns, Payne introduced some thoroughly modern practices, including voter registration drives, every-voter mailings, and active outreach to new citizens. The success of his methods earned Payne a reputation as "chief conjuror" of a startling turn-around in Milwaukee's Republican organization, and his influence soon extended to higher levels. By 1880 Payne was one of the acknowledged bosses of the state Republican Party and a national committeeman as well.

Although he never sought elective office, politics provided Payne with a livelihood. Patronage was the lubricant that kept party machines running, and Payne both received and dispensed his share of Republican largesse. He was appointed Milwaukee's postmaster by Ulysses S. Grant in 1876 and reappointed by Rutherford B. Hayes, James Garfield, and Chester Arthur; all four presidents, needless to say, were Republicans. Milwaukee's postmaster had more than 100 jobs to give away, and most went to party regulars. Payne was candid about his hiring policy. "I prefer giving employment to Republicans," he said, "other things being equal."

When Grover Cleveland, a Democrat, became president in 1885, Payne's tenure in the post office came to a sudden end. After more than a decade of service to his party, he had little trouble finding work. Payne parlayed his political contacts into a business career, earning a fortune as a lobbyist,

Henry Clay Payne, the politician and promoter who brought order to Milwaukee's chaotic lighting and traction scene.

director, and investor with ties to several companies, particularly firms that made or used electricity. The chaos prevailing in the Hinsey era made the situation ripe for an organizer. Payne had organizational skills to spare, but Milwaukee needed an entrepreneur who could bear the cost of consolidating its fractured utility system. Henry Payne formed an alliance with Henry Villard, and together they created the first truly metropolitan utility in the nation.

Henry Villard was a giant of the Gilded Age, and one of its least likely success stories. Born Ferdinand Heinrich Hilgard in 1835, he was the only son of a well-to-do Bavarian jurist. In 1853, when he and his father clashed over career choices, the teenager ran away to America. His intention, he wrote in his memoirs, was "to work hard there and to show his parents and sisters that he could make an honorable career for himself by his own endeavors." Ferdinand Hilgard became Henry Villard, but it took a long time for the immigrant to make a name for himself in other respects. Villard drifted around the country for years, seeking any employment that promised a rewarding future. He spent time in Cincinnati and near St. Louis — both German centers — and in 1856 migrated to Chicago, where he found a job selling encyclopedias of American literature. It was an odd choice for someone who had mastered English only recently, but Villard went at it with a will. His employer gave him "an entirely unexplored and very promising field" — the city of Milwaukee. Writing in the late 1800s, Villard recalled the city's allure to a young German:

Henry Villard, the immigrant tycoon who saw Milwaukee as the foundation of a national utility empire

Milwaukee has always been an almost German city. In 1856, the preponderance of the German element was even greater than at present; in fact, its Americanization, which has in the meantime progressed very rapidly, had then hardly begun. It was known among German-Americans as "Deutsch-Athen," and, comparatively speaking, deserved the name. There was a large number of educated and accomplished men among my countrymen, and in them the love of music and histrionic art was very marked.

Villard enjoyed his stay in Milwaukee, but he found the city's Germans far more interested in Goethe and Schiller than in Longfellow and Whittier. He sold only thirty-five sets of his literary encyclopedias in three weeks and left town five dollars in the red. "I had, as a rule," he wrote, "to drink continually of the cup of humiliation."

It was in Racine that Villard found what he considered, for a time, his vocation. Shortly after leaving the encyclopedia business, he was hired as editor of the Racine *Volksblatt*, a small Democratic paper that had just been purchased by German Republicans. Although the venture was a conspicuous failure, the *Volksblatt* launched Villard's career as a journalist. He moved to New York City soon after losing the paper. Working first as a freelancer for German weeklies and then as a correspondent for English-language dailies, Villard became a full-time newspaperman. His assignments took him across the country, and the young immigrant rubbed elbows with men like Abraham Lincoln and Horace Greeley. When the Civil War broke out, he filed dozens of stories from the front. Villard apparently found war reporting of more interest than his later financial dealings; Civil War stories take up nearly two-thirds of his memoirs.

Henry Villard was an established journalist by the late 1860s, but he was soon to become a power in the business world. The change in direction was almost accidental. Wealthy Germans had invested millions of dollars in American railroads, and they were troubled by the financial shocks that foreshadowed the Panic of 1873. On a brief return trip to his homeland, Villard was approached by a group of investors who desperately needed advice. His only qualifications were an ability to speak English, a thorough familiarity with the United States, and a keen native intelligence. Despite his lack of business credentials, the investors evidently preferred one of their own. They hired Villard as their agent in America.

His rise thereafter was meteoric. Villard proved an able guide to the intricacies of the railroad business, and he discovered a latent talent for "financiering." One successful deal in Oregon led to another in Kansas, and the investors' agent was promoted to railroad manager. A speculative frenzy had gripped the United States, and Villard was soon able to attract new money on the basis of reputation alone. In 1879 he formed a syndicate that purchased a group of steamship and railroad lines converging on Portland, Oregon. Villard controlled, or thought he controlled, the entire Columbia River watershed. The Northern Pacific Railroad was pushing west across the Great Plains at the time, and the financier was sure that its owners would have to come to him for an

Villard presided over the Northern Pacific's "last spike" ceremony in 1883.

outlet to the Pacific. When the Northern Pacific found an alternate route through his territory, Villard was outraged. With ample backing from capitalists in New York, Germany, and England, he formed the famous "blind pool," a group that purchased control of the Northern Pacific and made Villard its president. He was suddenly the head of an enterprise that employed 25,000 construction workers and was spending $4 million a month.

The Northern Pacific was completed between St. Paul and Portland in 1883. En route to the "last spike" ceremony, Villard led a picturesque entourage that included Ulysses S. Grant, several crowned heads of Europe, a gaggle of U.S. senators, and Chief Sitting Bull himself. The procession paused in Bismarck, where Villard laid the cornerstone of North Dakota's new capitol, and stopped again in Montana to watch 2,000 hired Crows dancing in full war regalia. An itinerant journalist only a decade before, Henry Villard was breathing the same rarefied air as the Vanderbilts, the Carnegies, and the Rockefellers.

The crash came quickly. Construction of the Northern Pacific had consumed a fortune, and the lands along its route were still too sparsely settled to generate adequate patronage. Shipments actually declined with the end of construction traffic, and the value of NP stock declined even more rapidly. Only three months after driving the last spike, Villard was practically insolvent. Control of the railroad passed to other hands. In his characteristic third-person style, Villard reflected on the reversal he'd suffered:

His fate was certainly tragic. Within a few years, he had risen from entire obscurity to the enviable position of one of the leaders of the material progress of our age.... But his fall from might to helplessness, from wealth to poverty, from public admiration to wide condemnation, was far more rapid than his rise, and his brief career was everywhere used to point a moral.

Villard was neither penniless nor beaten. After recuperating at his Hudson River estate for a few months, he left for a two-year sojourn in Europe, hoping that the change of scenery would help him forget "the falseness of friends and the outrageous vilification" of the American public. By 1886, when the lights were coming on in Wisconsin, he was ready to go back to work. Villard returned to the United States as the investment agent for two large German banks, and he had every intention of investing in railroads. He found, to his apparent surprise, that the struggling Northern Pacific was ripe for the picking. With $5 million from his "German friends," Villard acquired a controlling interest in the railroad and was soon directing its policy again. "What a revolution of the wheel of fortune!," he wrote.

Restored to his throne as a railroad king, Villard moved quickly to make the Northern Pacific a paying enterprise. His chief source of worry was Jim Hill's Great Northern Railroad, a new line pushing steadily west toward Puget Sound. As part of his effort to attract more traffic and to head off Hill, Villard decided to move the Northern Pacific's eastern terminus from St. Paul to Chicago. Instead of extending his own line through Wisconsin, however, he leased the track rights and properties of the Wisconsin Central Railroad, one of the state's mid-sized carriers. Villard also used the railroad as a blind to acquire depot properties and rights-of-way in Milwaukee and Chicago.

The Wisconsin Central was based in Milwaukee, and it was during negotiations there, presumably, that Henry Villard met Henry Payne. Which man initiated the discussions is unknown, but the pair evidently hit it off at once. The fruit of their early talks was a plan to consolidate Milwaukee's lighting plants and streetcar lines into a single utility, with Villard serving as capitalist and Payne as executive. Action followed quickly. In the autumn of 1888, at the midpoint of Boss Hinsey's reign as the do-nothing transit king, a group of New York brokers purchased the Milwaukee City Railway, the largest line in the state. The brokers were most likely acting as front men for Villard, for Henry Payne soon joined the railway's board. Although his role in the transaction was kept secret, the purchase of the Milwaukee City line was Villard's first step in the creation of a full-fledged electric utility.

Henry Villard operated at a scale that few Milwaukeeans could have imagined. His interests in transcontinental railroads, international banking, and a host of other fields made him a figure of global importance. Why, then, did this titan of finance back a scheme in a mid-sized city like Milwaukee? Because it served his larger interest in electrification. Villard, like Payne, was captivated by the emerging technologies of the electrical era. He had been among Thomas Edison's first backers; during his early days in Oregon, the promoter had outfitted one of his steamships with the first sea-going Edison lights in the world. In 1888, the same year he entered the Milwaukee market, Villard approached Edison with a novel proposition: the creation of a new company that would unite the inventor's diverse and increasingly disjointed sales and manufacturing firms. The outcome was the Edison General Electric Company, a conglomerate established in 1889. Villard, through his German associates, provided much of the new company's capital, and the immigrant financier became the firm's first president.

Villard's schemes for the electrical business were every bit as ambitious as his visions of rail monopolies. He had begun to dream of a utility empire. Villard saw Edison GE as a gigantic holding company, owning and operating central stations that would power streetcars, lights, and machinery in cities from coast to coast. Forming captive utilities, he believed, would be more profitable than selling equipment to a horde of disorganized independents. Villard was on intimate terms with some of Europe's leading investors and inventors, and he may have dreamed of a truly international cartel, a combine that would control the flow of power in every country on earth. The Milwaukee system was a minuscule part of a grandiose scheme, but it was also, in a sense, Henry Villard's flagship. His Milwaukee ventures represented Villard's first foray into the utility business, and they became, in the end, the most fully developed of all his electrical properties.

With their master plan in place, the two Henrys proceeded methodically to wipe out the competition that had plunged the city's lighting and transit

JULY 22, 1890.

VILLARD IS IN IT

Cream City Road Controlled by Him.

THE REPORTED TRANSFER CONFIRMED YESTERDAY.

The Line Is to Be Equipped With the Sprague System, the Villard Syndicate Taking Pay for the Work in Bonds.

The transfer of the Cream City railroad to the syndicate which owns the Milwaukee City lines, announced by The Sentinel some ago, was confirmed officially yesterday. It was not a sale in the narrow sense of the word, but a partial transfer of stock. The Edison-Villard syndicate is now in control of all the electric railway patents in America. It controls both the Thompson & Houston company and the Sprague company. The owners of the Cream City lines have been exerting themselves for the last two months, trying by all means to secure money to pay for the electric equip-

After months of rumors, the Milwaukee Sentinel *revealed Henry Villard as the man behind Milwaukee's streetcar consolidations.*

service into chaos. The Milwaukee City Railway purchase was consummated by the end of 1888. In July, 1889, the Edison Electric Illuminating Company of Milwaukee was organized, with Henry Clay Payne as its vice-president. (The company, as noted in Chapter 1, received a city franchise in September, breaking S.S. Badger's monopoly on Milwaukee's lighting business.) In the spring of 1890, a Pittsburgh syndicate purchased the Cream City Railroad; Henry Clay Payne became its president. The Pittsburgh group, it was rumored, was acting as a front for Henry Villard, and his frequent visits to the city only added fuel to the fire.

The secret was out before long. "Villard Is In It," the *Milwaukee Sentinel* finally announced on July 22, 1890. The wonder of Wall Street was declared the power behind both the Cream City and the Milwaukee City purchases. It was as if Donald Trump had entered the Milwaukee real estate market, and the news was greeted with a predictable mixture of awe, disbelief and, in some quarters, suspicion. Henry Payne, whom the *Sentinel* identified as Villard's local agent, was careful to focus on the future in an August interview. He promised, among other things, a quick end to the transit mess:

> *I suppose that I will have the general management of these roads, and I am bound to see that they are absolutely first-class in every way. The people of Milwaukee deserve the best that the world affords and I am determined that they shall have it in electric street railways as far as our lines are concerned. Mr. Villard, who has a controlling interest in these lines, has just gone to Europe. His final instructions before taking his departure were to make these Milwaukee roads as good as the best.*

The newspaper reported only the half of it. Villard, working with and through Payne, wanted a monopoly; he wanted complete control of every aspect of the city's electrical business. From the time electricity was introduced in Wisconsin, lighting and traction — the loads that could keep generators humming night and day — had traveled separate but parallel paths. Relatively small operators like S.S. Badger didn't, or couldn't, see their essential unity, but visionary capitalists like Villard did as a matter of course. At his instigation, the two paths were about to converge.

The North American Company

The holding company was among the favorite tools of Gilded Age businessmen. It allowed them to direct capital to a multitude of enterprises with a maximum of control and a minimum of disclosure. Henry Villard's key holding company was the Oregon and Transcontinental, the legal parent of the Northern Pacific Railroad. A massive debt restructuring rendered the railroad independent in 1889, and Villard formed a new company to channel funds to his smaller holdings, including the Milwaukee purchases. The North American Company was incorporated in New Jersey on June 14, 1890.

Although its charter gave North American the right to enter practically every business but fruit-peddling, the company's specialty was electric utilities, a field it shared with Edison General Electric. The relationship between the two companies was ambiguous. Villard was the ultimate head of both, and their officers worked cooperatively in several markets, but key Edison executives, including Thomas Edison himself, resisted their president's enthusiasm for the central-station business. They were more than willing to manufacture and sell electrical systems, but operating those systems was quite another matter. It is likely that, in forming North American, Villard was simply hedging his bets. If Edison GE declined to build a central-station empire, Villard, through North American, would do it himself, and he would start in Milwaukee.

The new company got off to a troubled start. Bowing to pressure from farmers, miners, and laborites, all of whom demanded a more liberal monetary policy, the federal government was moving from the gold standard to a money supply based on silver as well as gold. The short-term result was a devalued currency, which prompted investors to hoard their gold and stay out of the financial markets. North American came into being with debt totaling millions of dollars, and the company continued to borrow heavily to finance its acquisitions. In late 1890, when the first of its notes came due, a tight money market and general shakiness on Wall Street made refinancing impossible. North American was forced to surrender some of the securities it had offered as collateral. Villard, who was still on the European journey mentioned by Payne, sent $2 million supplied by his German backers, but the situation worsened. Decker Howell, the New York brokerage house that carried the lion's share of North American's bonds, failed in November. Villard returned to America to find that the company's creditors had taken most of North American's holdings to satisfy their debts. "It had been stripped," he lamented, "by the forced sales of the great bulk of its assets at a heavy loss, and was prostrate and reduced to inactivity for years to come."

Two of the largest holdings that escaped liquidation were electrical properties in Milwaukee and Cincinnati. Both cities, interestingly, were major centers of German settlement, and both had figured prominently in Villard's introduction to the New World. Despite his well-deserved reputation for ruthlessness, there is the intriguing possibility that Villard held on to the properties for old time's sake.

North American may have been "reduced to inactivity" by the turmoil on Wall Street, but inactivity is relative. The Milwaukee operation, by far the company's largest asset after the fall, was practically untouched by the travails of its parent. As a sign of Villard's commitment to the project, the flow of capital remained steady, and the work of consolidation continued. North American formed a subsidiary holding company to manage the venture: the Milwaukee Street Railway Company, incorporated on December 22, 1890. Henry Payne shuttled between New York and Milwaukee to work out the details, spending seven of 10 consecutive nights on sleeper cars. Henry Villard was MSR's president, but Payne, as vice-president, was responsible for day-to-day management.

The Milwaukee manager's first task was to gain control of the

The North American Company was formed in 1890 to provide a "basket" for all of Villard's utility ventures.

odds and ends that constituted the city's lighting and transit service. It took well over a year. With the Milwaukee City and Cream City lines as a foundation, the Milwaukee Street Railway struck deals with everyone on the scene. Badger Illuminating, the city's pioneer utility, was among the first acquisitions; Edison Electric Illuminating and Milwaukee Electric Light followed it into the fold. (The Edison purchase signaled Villard's intent to operate independent of Edison GE.) Boss Hinsey and Charles Pfister sold their two lines — the Milwaukee Electric Railway and the Whitefish Bay road — in the summer of 1891, and Washington Becker agreed to a deal for his West Side line a few months later.

The Milwaukee Street Railway undoubtedly paid a premium for the operating companies it acquired; would-be monopolists seldom find bargains. (The Hinsey line, not surprisingly, was the least expensive railway acquisition.) The situation was complicated by the presence of phantom companies. Two factors — the Common Council's open-door policy to franchise-seekers and the obvious presence of a shark in the waters — led to a speculative frenzy in Milwaukee.

An early example of indoor lighting at Milwaukee's Industrial Exposition Building

The Milwaukee Street Railway, a North American subsidiary, absorbed all its competitors to become the first unified utility system in the nation.

Small companies were formed with every intention of being eaten — for a price. The *Milwaukee Sentinel* (July 29, 1890) described the climate:

> *There is a craze at present in Milwaukee for street railway franchises. Real estate men, for purposes of speculation, are conceiving all sorts of schemes, and soon there will be hardly a street in the city not taken up by railway tracks if all the franchises asked for are granted by the council.*

Newcomers sprang up in the lighting field as well, and the Milwaukee Street Railway was forced to deal with all of them. Although the buying activity continued for several years, consolidation was substantially complete by the end of 1891. The price tag, in cash and securities, exceeded $5 million — more than $85 million in 1996 dollars.

Henry Payne demonstrated considerable skill in assembling the components of a system. (It was a mark of his, and Milwaukee's, importance to Villard that Payne was one of only two Midwesterners on the North American board.) He was a deal-maker, a cool-headed negotiator who could size up the issues and come to terms with an absolute minimum of fuss. Payne put together the Milwaukee system with the same shrewd pragmatism he showed in building political coalitions. As later events would demonstrate, his forte was not public relations. Payne's skills were better suited to the back room than to the public arena, and he made few efforts to project an image of warmth and human concern. "I have never been a victim of procrastination," he once told a reporter, "and I have never waited for other people to do my work. If you have anything to do, do it." The *Milwaukee Journal* (Oct. 5, 1895) offered a frank assessment of Payne's character, calling him an "able man with no music in his make-up and no poetry in his soul — a grasper of opportunities, an amiable master of material things, a plain, unassuming Yankee everlastingly saturated with common sense."

The pragmatic Yankee did not assemble Milwaukee's utility system alone. One of his chief lieutenants in the effort was Edward C. Wall, who, as the long-time president of Badger Illuminating, had charge of the lighting acquisitions. The two executives seemed unlikely bedfellows at first glance; Payne was still a Republican boss in 1890, and Wall was head of the state's Democratic Party. Their alliance underlined the easy mingling of business and politics that marked the Gilded Age. "Both have been called Napoleons," wrote a reporter in 1895, "and not without warrant; for both are men of high intellectual powers, magnificent dash and daring, and fine strategic skills." The two field generals had held similar patronage jobs (Payne as postmaster, Wall as revenue collector), they maintained offices in the

Utility poles and webs of wire became common sights in the region's cities.

same building, and they reportedly bent elbows over the same poker table. Payne and Wall may have been spirited adversaries at election time, but the business of America in the Gilded Age was business, regardless of party affiliation.

With the final consolidation of the city's streetcar and lighting companies, an era came to an end. One by one, the old protagonists moved on to other pursuits. Sheridan S. Badger, still based in Chicago's Loop, turned his attention to mining investments. Washington Becker, after returning to the practice of law for a few years, found work as president of the Marine Bank. Charles Pfister, John Hinsey's financial angel, became one of MSR's most prominent local investors. Boss Hinsey himself moved to Chicago in 1894 and retired to California in 1910. With brutal irony, Milwaukee's one-time transit king was killed by a Los Angeles streetcar on April 30, 1911.

Rising, Falling, Rising Again

With consolidation complete, the Milwaukee Street Railway undertook a considerably more imposing task: turning the sundry bits and pieces it had acquired into one cohesive utility. What Villard and his backers had purchased was the *right* to build a monopoly, not a working system, and they faced some daunting obstacles. The equipment inherited by the company was hopelessly mismatched; the West Side line, for instance, had Sprague cars, while the Hinsey line used a Thomson-Houston system. Some acquisitions were, in the apt phrase of Henry Payne's biographer, "little more than a franchise and a streak of rust." The entire system was full of redundancies; arc light service was available from two central stations, and streetcar routes overlapped in every section of the city. As head of Edison GE, Henry Villard was able to supply equipment at cost, but the expense of electrification was nonetheless staggering. Payne and his Milwaukee staff had to build a system essentially from scratch.

In contrast to the torpor of the years just previous, MSR's pace was dizzying. Edward Wall supervised dramatic expansion on the lighting side. The old Badger facility on the Milwaukee River at Vliet became the city's sole arc lighting plant, and new generators increased its service from 400 lights in 1890 to 1,500 in 1893. They provided illumination for more than 40 miles of city streets. A few blocks down the river, near Oneida (now East Wells) Street, the brand-new Edison plant was supplying power for 13,000 incandescent lamps by 1893. Its boilers were rated at 8,000 horsepower (compared to Badger's 900), making the Edison

The Edison plant on the Milwaukee River was among the largest in the country — relatively speaking. It supplied current for a grand total of 13,000 incandescent lamps in 1893.

Electric streetcars became a fixture in the urban landscape. Car 256 served the Farwell Avenue line . . . (Ed Wilkommen)

facility one of the largest power plants in the country. Electric lamps replaced gaslights in hundreds of shops, offices, and private residences, and the plant was beginning to drive machinery in the city's growing industries as well. By 1896 the Badger and Edison plants had a combined generating capacity of 4,150 kilowatts — an infinitesimal fraction of Wisconsin Electric's current 5.6 million kilowatts, but a most impressive figure in the late nineteenth century.

Much of the Edison plant's capacity was used to power streetcars, and the city's transit system grew even more rapidly than its lighting network. Henry Payne led the effort to rationalize a thoroughly irrational system. He pruned unprofitable routes (often incurring the wrath of local residents), combined overlapping routes that had survived from the horsecar era, purchased the latest in rolling stock, and extended service in virtually every direction. The Milwaukee Street Railway network grew to 16 separate lines, and all were electrified as rapidly as possible. (Veteran horsecar drivers welcomed the change, but some couldn't help saying "Whoa!" every time they came to a stop.) The longest one-fare ride in the city increased from 4.25 miles in 1890 to 9.25 miles in 1896, and the system's total trackage soared from 53 miles to more than 130 during the same years.

The work of consolidation and conversion was not entirely trouble-free. Although the transit system soon covered more ground than all its predecessors combined, city fare remained one nickel — a definite drag on the firm's earning power. An 1892 fire destroyed the Kinnickinnic Avenue carbarn; nearly all the leftover horses and mules were saved, but more than 100 cars (some of them brand-new Sprague models) went up in smoke. The company also faced political pressures. The marked improvements in transit and lighting service were obvious to everyone, but some citizens, notably the reformers on the post-Hinsey Common Council, were troubled by the monopoly status they had granted to MSR. The aldermen tinkered with the company's fare structure, adopted new safety regulations, and revived the old campaign for municipal ownership.

. . . while Car 258 carried riders out to the old West Side Street Railway's terminus at 36th and Wells.

Payne and his associates cleared the hurdles with relative ease, and the work of system-building continued. The result was a startling turnaround. Victimized by political intrigue and their own lack of vision, Milwaukee's transit and lighting companies had been decidedly behind the times for years. The Milwaukee Street Railway struck like a lightning bolt. With an abundant supply of capital and just as much energy, its leaders built a model electric utility. The national significance of the Payne-Villard venture should not be overlooked. Just as North American was the first of dozens of holding companies that would later dominate America's utility industry, its Milwaukee project was the first unified utility system in urban America. Once a backwater in the emerging electrical world, Milwaukee took its place in the vanguard of the nation's cities.

Local bystanders could only watch in amazement. Winfield Smith, former head of the Cream City line, penned a faintly wistful description of the city's transit monopoly in 1892:

> *Consolidation of companies has destroyed competition, and given almost unrestricted choice of lines, courses and distances. Improvements of various kinds, including the construction of cars unsurpassed in any shops in this country, are under way, and the managers of the Milwaukee street railroads appreciate and seize all their opportunities.*

One clear outcome of the Milwaukee Street Railway's expansion effort was a real estate boom. Payne's stated purpose was to bring the entire city within 30 minutes of downtown, and he made substantial progress toward that goal. Developers of the time began to advertise their properties as "within four blocks of streetcars" or "walking distance to terminus of electric line." Real estate sales in the city jumped from $5 million in 1887 to more than $10 million in 1892 — a record that stood until after World War I. Milwaukee remained a densely settled community, at least by American standards, but the streetcar played a major role in dispersing the city's population.

It even appeared, for a time, that the long-cherished dream of suburban homes for the workingman would become a reality. The *Milwaukee Sentinel* (Apr. 24, 1892) restated the vision:

> *Rapid transit enables a would-be buyer and house builder to go further away from the center of the city where he would have to content himself with a dingy little house and 30-foot lot, while he can build to his heart's content, with floods of light and health-giving sunshine and a fine sward in the bargain.*

MSR did eventually extend service to suburbs like Cudahy, South Milwaukee, and North Milwaukee. Those, however, were industrial suburbs, and they attracted a resident work force, not commuters. More often than not, transplanted workers found the same densities they had known in the city. In the more purely residential communities — Whitefish Bay, Wauwatosa, Tippecanoe, and others — it was relatively affluent commuters who took the streetcar to "floods of light and health-giving sunshine."

Henry Payne developed North Milwaukee as an industrial suburb and named its main street, Villard Avenue, for his boss. The original village hall (left) is still standing.

Developers who wanted streetcar service to outlying areas, whether their projects were industrial or residential, generally paid for it; construction subsidies were an important source of revenue for most urban transit systems. In the case of North Milwaukee, no subsidy was necessary, for the suburb was one of Henry Payne's extracurricular activities. Its site (near 35th and Hampton) lay at the intersection of two major Milwaukee Road lines, making it a prime spot for industrial development. Payne, who had previous experience as a townsite promoter for the Milwaukee Road, purchased large tracts of land in the area and launched an aggressive sales campaign. In the early 1890s, he formed development companies, courted industrial buyers, and sold lots to all comers. Payne also extended streetcar service to the fledgling community, crossing nearly two miles of open land, and North Milwaukee received its village charter in 1897. As the suburb's leading developer, Payne named its main street for his boss, Henry Villard, and the name survived when Milwaukee annexed the community in 1929. Villard Avenue today is the city's only concrete reminder of the financier who created the utility known as Wisconsin Electric Power.

Villard himself watched the system's growth with interest, but he was not an active participant; daily management was left to North American's officers and Henry Payne's staff. Villard, as always, devoted himself to larger interests, particularly his dream of a national electrical monopoly. Events in Milwaukee were, in fact, a microcosm of consolidations taking place on the national level. Soon after becoming Edison General Electric's president in 1889, Villard had brought the Sprague Electric Railway Company into his firm, giving Edison GE the most advanced transit technology in the world. Thomson-Houston, the Edison conglomerate's largest competitor, was just as aggressive, buying the Van Depoele traction firm in 1888 and Charles Brush's arc-light company a year later.

Three giants now dominated America's electrical industry — Edison GE, Thomson-Houston, and Westinghouse — and Villard resolved to bring the three together under his own leadership. He first courted Westinghouse, the smallest of the trio, but Villard soon decided to reach for the brass ring. In 1891 he began negotiations with Charles Coffin, head of Thomson-Houston, and it appeared that a deal was imminent. J.P. Morgan, however, was persuaded to finance the transaction on terms that gave Thomson-Houston the dominant position. The present General Electric Company was formed by merger on June 1, 1892, and Henry Villard found himself excluded from its management.

The formation of General Electric effectively ended Villard's dreams of a worldwide electrical trust. He had lost control of a giant, and his own North American Company, crippled by its creditors in 1890, was in no position to compete in the national arena. The financier did not abandon the seeds he had planted in Milwaukee, but he was soon tilling another field. Villard was profoundly disturbed by a continuing series of tremors in the financial markets, tremors that he blamed on America's slackening ties to the gold standard. Characteristically, he took it upon himself to lead the federal government back to a currency based on gold. Villard was so absorbed in lobbying, he wrote, that he felt his business responsibilities "but lightly." In an effort to cut back his corporate involvements, he resigned the presidency of the Milwaukee Street Railway in 1892.

As Villard stepped into the background, Henry Payne's role become more central. His pace in the early 1890s must have been frenetic. In addition to managing the Milwaukee Street Railway, Payne was a nationally prominent Republican leader, president of Wisconsin Telephone and, surprisingly, a director of the Milwaukee Gas Light Company, among his other corporate commitments. He also found time for outside activities like the North Milwaukee development scheme. Payne stood alone at the helm of MSR until 1893, when he hired C. Densmore Wyman as general manager. Wyman later gave a full measure of credit to his boss:

> *He was a tower of strength and stimulation to his subordinates, and it was largely due to his unswerving faith in the future development of the company's business that the enterprise was sustained and ultimately established upon a profitable basis.*

More concrete recognition of Payne's work came in the same year he hired Wyman. Members of the American Street Railway Association, during the trade group's 1893 convention in Milwaukee, elected Henry Clay Payne their president.

By the summer of 1893, the Milwaukee Street Railway's system was substantially complete. Although the last horsecar was not retired until 1894, electrified transit was an accomplished fact, and electric lights were turning night to day throughout the city. "Our streets are brilliantly lighted at all hours," wrote one local booster, and another called the streetcar system "one of the best in the United States." The work had cost the North American Company millions of dollars, but MSR was beginning to earn money for its parent. Henry Payne and, from a greater distance, Henry Villard could look on the system they had built with justifiable pride.

They did not have long to enjoy the view. Just as Villard had feared, a catastrophic chain reaction ripped through the financial markets in the summer of

As the system grew, more and more troubleshooters were needed to keep the current flowing.

1893. As the government's flirtation with the silver standard continued, more investors called their loans and hoarded their gold. As the money supply contracted, businesses were unable to repay their debts, much less expand. Companies in all fields began to fail, and one failure led to another with devastating speed. The Panic of 1893 brought a long period of prosperity to an end, and the weeks of panic gave way to more than two years of depression. By the end of 1893, more than 600 banks had failed, and a quarter of the nation's heavy industries had closed their doors. Railroads covering 22,500 miles of track were in bankruptcy, and bumper crops lay rotting in the fields.

For the Milwaukee Street Railway, the depression's timing could not have been worse. As Milwaukee's unemployment rate soared, ridership on the new streetcars plummeted, and demand for electric lighting evaporated. MSR's revenues for June of 1894 were half the previous June's. The company's expansion program came to an abrupt halt, and Henry Payne adopted a new policy of retrenchment. Despite desperate attempts to cut expenses, a tide of red ink continued to rise around the system he had built. By the middle of 1894, the Milwaukee Street Railway was unable to pay the interest on its considerable debt. By the middle of 1895, the company was bankrupt.

Payne's boss, in the meantime, had retired from the fray. Sensing danger, Henry Villard had distanced himself from his manifold business interests. He left the boards of both the Northern Pacific Railroad and the North American Company only weeks before the crash of 1893. The Northern Pacific quickly sank into bankruptcy, and Villard's last act on its behalf was to organize a committee of receivers. (One of his choices was Henry Clay Payne.) Villard retained a financial interest in North American and other old flames, but his business career, for all practical purposes, was over. That career coincided neatly with the boom times between the Panics of 1873 and 1893. His fortune secure and his appetite for wheeling and dealing fully satisfied, the financier retired to enjoy what he called an "abundance of leisure ... for extensive reading, literary labor, reform work, and the philosophic contemplation of the momentous events following each other so rapidly in our time."

The North American Company, battered in 1890, was barely breathing after the Panic of 1893. Its principal asset, in fact, was the bankrupt Milwaukee system. Henry Villard may have envisioned his Milwaukee operation as the cornerstone a national utility empire, but it became, in the twilight of his career, an isolated remnant of a grandiose dream.

However reduced in scope and resources, North American remained in business, and the holding company's survival was the Milwaukee system's salvation. By 1895 MSR had defaulted on all its bonds, but the firm's sole bondholder was North American. Fully aware that few investors were willing to gamble on its pioneering Milwaukee project, the parent company had kept all of the Street Railway's securities in its own treasury, hoping to sell them when the

There was still room for horses: an early version of the line "truck."

system was sufficiently developed to attract outside investors. If the stocks and bonds had been sold outside the corporate family before 1893, there is every likelihood that the bankrupt Milwaukee system would have been dismembered in a sheriff's sale. Instead, MSR was like an errant teenager whose parents stood ready to bail him out. As sole creditor, North American had the right to appoint the firm's receivers, reorganize its business affairs, and refinance its debt. In May, 1895, the company began to exercise that right.

There must have been some anxious moments, but it seems clear that no one at either MSR or North American seriously considered dissolving the system. Henry Payne was one of the Milwaukee Street Railway's receivers, and he continued to operate the utility while the work of restructuring proceeded. As business conditions slowly returned to normal, Payne began a cautious resumption of MSR's expansion program. A long-promised line to Cudahy was completed by the autumn of 1895, when the company was still in receivership.

From the ashes of the Milwaukee Street Railway, finally, rose a new enterprise — The Milwaukee Electric Railway and Light Company. It was incorporated on January 29, 1896, with William Nelson Cromwell, a principal in North American, as TMER&L's president and Henry Payne as the firm's vice-president and resident officer. North American issued $14 million in stocks and bonds for its new subsidiary, and the financial markets, fully recovered by now, absorbed the TMER&L offering quickly. The crisis was over, and the reincarnated utility resumed the work of system-building with new energy.

TMER&L was perhaps Henry Villard's most enduring legacy. Practically all of his other projects passed into other hands, but the Milwaukee system continued down the path he had chosen for it in the beginning. Shortly before his death in 1900, Villard finished the memoirs he had been laboring over since retirement. Writing in the third person, the immigrant captain of industry described his Milwaukee venture with undisguised pride:

One of his transactions was the acquisition of all the street railway lines in Milwaukee, their change from animal to electric power, and their consolidation with the local electric lighting interests into one corporation, resulting, for the first time in the United States, in the distribution of electrical energy for light, power, and traction purposes from one central station. This combination has since grown into one of the largest and most successful light, traction, and power companies in the country.

That company was TMER&L. It would rise from the tangled roots of the formative years — the era of horsecars and arc lights, of Henry Jacobs and Boss Hinsey — to a position of leadership among Midwestern utilities. It would build, and later dismantle, one of the finest electric transit networks in the nation. It would branch out from its Milwaukee base to serve a region covering more than 12,000 square miles. It would, finally, 40 years after its founding, take on new life under a new name — the Wisconsin Electric Power Company.

The Milwaukee Electric Railway and Light Company (TMER&L) marked a fresh start for utility service in the region. This one proved successful.

Chapter 3

1896-1911

Mr. Beggs Builds an Empire

The new company's initials said it all. As "TMER&L Co." was emblazoned on timetables, transfers, and a growing number of buildings, the firm proclaimed itself *The* Milwaukee Electric Railway and Light Company. Its insistence on "The" underlined TMER&L's hard-won status as the sole supplier of power in Wisconsin's largest city, brooking no opposition and tolerating no competition to its monopoly. As "TM," the shorthand acronym, gained wide acceptance, the point was made even more boldly.

The company's position as *the* local utility had not been easily achieved, and it was even less easily maintained. The years between 1896 and 1911 were some of the most expansive in the company's history, but they were also some of the most difficult. TMER&L grew with remarkable speed during the period, spreading out from its Milwaukee base to blanket the region with transit lines and power corridors. The presiding genius of that growth was John I. Beggs, a relative newcomer who earned an enduring reputation as a penny-pinching autocrat of extraordinary effectiveness.

During the same years, however, TMER&L was at the very center of a political firestorm. The same depression that crushed the Milwaukee Street Railway after the Panic of 1893 gave rise to an ambitious reform movement, and its favorite target was the Street Railway's reincarnation. Dominated by Progressives on the state level and socialists in Milwaukee, the reform coalition took every opportunity to make life difficult for TMER&L. Whether fighting a bitter strike in 1896 or facing the election of a socialist mayor in 1910, the company was constantly on the defensive.

John Beggs directed aggressive expansion on the one hand and defensive maneuvers on the other — a blend that made the period between 1896 and 1911 one of the most schizophrenic in the utility's history. TMER&L's development proceeded in tandem with endless public commotion; John Beggs and his associates sometimes divided their days between planning new lines and making court appearances. Both sides of the story — the growth of a system and the rise of public protest — are told in the two principal sections of this chapter.

I. The Beggs Era

Henry Payne remained at the helm, at least nominally, for six years after TMER&L was organized. As the company's vice-president and resident officer, Payne was the system's most visible figure in the community, but he occupied an intermediate position. Operating details were left to TM's general manager, C. Densmore Wyman,

John Irvin Beggs, the administrative genius who guided TMER&L's expansion with a tight fist and an iron will

who had complete charge of the rolling stock, power plant equipment, and employment. Larger policy issues were the province of the North American Company's board. Although Payne remained an influential board member, control of the holding company had passed from Henry Villard to a well-heeled group of New York bankers and attorneys. North American's president in 1896 was Charles W. Wetmore, a Harvard-trained lawyer who had reorganized the firm after the Panic of 1893. Widely known as a yachtsman, Wetmore commuted to Manhattan from a Tudor mansion overlooking Oyster Bay on Long Island. The company's treasurer, and in 1902 its board chairman, was George R. Sheldon, another Harvard alumnus. Sheldon was a prominent banker who happened to serve with Henry Payne on the Republican National Committee. The veteran of the group was attorney William Nelson Cromwell, one of Villard's cronies from his railroading days. With his shaggy white mane and Buffalo Bill mustache, Cromwell may not have looked the part of a Wall Street lawyer, but he was one of the most highly regarded corporate attorneys in the nation. His firm did North American's legal work, and Cromwell himself served as president of The Milwaukee Electric Railway and Light Company until 1902. His TM post was largely ceremonial, but Cromwell was Henry Payne's immediate boss.

North American was a part-time job for Cromwell, Wetmore, Sheldon, and the other men who controlled its affairs. Every one of them had a multitude of other interests, but the directors took their responsibilities seriously. As the American economy

William Nelson Cromwell, the New York lawyer who served as TM's part-time president

pulled out of the depression of the mid-1890s, they began, slowly and cautiously, to pursue Henry Villard's dream of a national utility empire. The Milwaukee system remained North American's flagship, but Villard's corporate heirs retained their Cincinnati interests and purchased new holdings in Detroit and St. Louis. The holding company adopted a fluid, vaguely federal style in managing all its operating units. Although the major decisions were made in New York, local managers were allowed a great deal of latitude, and major local investors were generally granted seats on the parent company's board. By 1906, for instance, five of North American's 16 directors were Milwaukee residents.

The operating subsidiaries also furnished a pool of executive talent that flowed easily from one part of North American's system to another. The most notable member of that pool was, without question, John Irvin Beggs. Beggs was, in fact, one of the most singular characters in the history of the American utility industry, and the story of his rise to power and wealth is the American Dream incarnate. Born in Philadelphia in 1847, Beggs entered the work force somewhat earlier than most of his peers. When his father died, leaving Beggs and his mother penniless, the boy took a job in a local brickyard. He was seven years old. As other jobs followed — cattle drover, carpenter's helper, butcher's boy — Beggs, not surprisingly, began to harbor dreams of wearing a celluloid collar. He worked his way through business school in Philadelphia, showing such aptitude for accounting that he was hired as an instructor after graduation. He soon gave up the classroom for the business world. At the age of 21, Beggs moved to Harrisburg, Pennsylvania's capital, to work as an accountant for a coal company, at the magnificent salary of 10 dollars a week. His next step was self-employment. Beggs opened a real estate and insurance office that became, over the years, one of the most prosperous in Harrisburg. The entrepreneur put down roots in the city, marrying a local girl, Sue Charles, and raising a daughter, Mary Grace.

Beggs lived in Harrisburg for nearly

20 years, growing into roles as a leading businessman, an active Mason, and a pillar of Grace Methodist Church. It was church work, fittingly, that led to his true calling. In 1883, when Grace Methodist was planning a new building, Beggs took charge of the project. "I used to tell the pastor and the members," he recalled in 1911, "that if they would look after the spiritual end that I would look after the temporal end." The new church, Beggs decided, would have the very latest in lighting: the incandescent lamps that had flickered on in Thomas Edison's Pearl Street Station only a year before. It dawned on Beggs that a complete central station, serving customers throughout Harrisburg, had definite business potential, and he entered a new field as the financier, builder, and manager of one of the first Edison plants in the country. Successful from the start, the Harrisburg system launched Beggs' real career. In 1887, shortly before his fortieth birthday, he accepted Thomas Edison's offer to take charge of the lighting system in New York City. There he met and befriended other giants, including Henry Villard and Samuel Insull.

Only three years later, Beggs became Edison General Electric's western manager, supervising the company's Midwestern affiliates from a new office in Chicago. Although he never severed his Harrisburg ties (Beggs is buried in the Grace Methodist cemetery), the Midwest was his home for the rest of his life. Wisconsin was part of the circuit; Beggs served as a director of the Edison affiliate in Appleton, and he developed a working relationship with Henry Payne and other utility leaders in Milwaukee. He also developed a

John Beggs as a young businessman, probably in Harrisburg, Pennsylvania

national reputation. Beggs served seven consecutive terms (from 1886 through 1892) as president of the Association of Edison Illuminating Companies, a trade group comprising all the pioneer Edison central stations.

The 1892 merger that created the General Electric Company took Beggs' career in other directions. With the Thomson-Houston faction in control of GE and the Edison-Villard faction out in the cold, Beggs lost his job. He found another one, and quickly. Villard's North American Company hired him as the general manager of its subsidiary in Cincinnati, and there he stayed through the depression of the mid-1890s. When the economy recovered, North American relied heavily on Beggs' counsel in shaping its other ventures, including the Milwaukee system. In February, 1896, Beggs was retained as a part-time "expert adviser" (at $3,000 a year) to direct a power plant expansion for TMER&L.

Several months later, C. Densmore Wyman resigned as TM's general manager to take a similar position in New Orleans. (The strike described later in this chapter undoubtedly played a role in his decision.) The Milwaukee system needed major assistance, and North American turned instinctively to Beggs. He recalled that, on one of his regular trips to the corporate offices in New York, "William Nelson Cromwell, the lawyer, ... jumped me." "Out in Milwaukee," said Cromwell, "the streetcar situation is in a muddle. Go out there and straighten things out." Beggs made a thorough inspection of the system and returned to New York with, as he put it, "a hatful of ideas." The North American board promptly endorsed the whole hatful and challenged Beggs to make them realities. On April 1, 1897, John I. Beggs began his long tenure as general manager of TMER&L. "I didn't want the job," he said years later, "but they got my Scotch-Irish up."

The new executive doubled as manager of the Cincinnati lighting utility at first, but North American soon disposed of its interest in that firm, freeing Beggs to devote all his energy to the Milwaukee operation. North American rewarded his efforts handsomely. In 1902 Beggs succeeded William Nelson Cromwell as president of TMER&L; he received an annual salary of $15,000 — roughly $270,000 in 1996 dollars — with bonuses of preferred stock worth several times that figure.

The good news was followed quickly by bad: Sue Beggs died on March 14,

1902, only a month after her husband's promotion. Work had always been Beggs' primary focus in life. With his wife gone and his daughter on her way to adulthood, work now became practically his only focus, and the pace of TMER&L's expansion program accelerated. In 1909 Beggs' salary was doubled to $30,000, and it is likely that North American was getting a bargain.

Beggs the Builder

John Beggs professed to find nothing but chaos when he took over the Milwaukee system in 1897. "When I came here," he recalled, "I found 300 single-truck cars and a system of antiquated, almost worthless tracks.... It looked dark when I came here. It looked like the impossible." Never one to downplay his own accomplishments, Beggs could not resist understating those of his predecessors.

Henry Payne had not, in fact, been sleeping since 1890. He and his associates, notably Densmore Wyman, had taken bits and pieces of overlapping companies with mismatched equipment and molded them into one coherent system. The tangible results were dramatically improved transit service and a significantly brighter city after sundown. In 1895, only five years after the first electric streetcar clattered down Wells Street, the Milwaukee Street Railway carried 28 million riders, and the use of electric power was becoming general. The *Milwaukee Sentinel* (Mar. 17, 1895) described the prevailing trends:

> *Every street car nowadays is lighted by electricity, every office has a telephone, every building is a net of electric wires. Electricity forces its way into everything and the day has already come when it has begun to supplant fuel.*

Available for the magnificent sum of five dollars an hour, the Marguerite *was a party on wheels in the Gay Nineties.* (Russell Schultz)

Nor had Payne ignored opportunities to build the system. He opened suburban lines, some of which blossomed into interurban routes, and took imaginative steps to increase city ridership, particularly on Sunday, the deadest day of the week for transit companies. In 1895 he began to offer free band concerts in both Lake and Washington Parks as an inducement to weekend travel. In 1896, months before Beggs' arrival, he unveiled the *Marguerite,* a party car that was rented to revelers for five dollars an hour. Outfitted with 700 incandescent lights, wall-to-wall carpeting, and "obliging porter service," the *Marguerite* was a picturesque sight on Milwaukee's streets during the summer months.

Henry Payne left an indelible mark on the system, but Payne and Beggs were different people with significantly different tasks. Payne's primary task was consolidation: transforming the chaos he had inherited into a single system. Beggs' principal focus was expansion: growing the system he had inherited into a regional utility empire. He repeated, on a larger scale and at a higher level, the work Payne had undertaken in Milwaukee between 1890 and 1896. Although Beggs eclipsed Payne as the driving force behind TMER&L, both men played essential roles.

John Beggs was admirably fitted, both by temperament and by talents, to the role of empire-builder. He was, first of all, an absolutely tireless worker. When a reporter asked the secret of his success, Beggs said, "Never fret. Work hard, persevere, and have endurance. Hard work, if the work be congenial, has never killed or harmed any person." No one practiced those precepts more faithfully than John I. Beggs. In a career that began at the age of seven, he worked, in his own phrase, "30 hours a day and nine days a week" until he was literally on his deathbed. Sunday staff meetings were routine during his tenure.

Beggs was also a man of legendary frugality, a legacy, no doubt, of his early struggles. When he ordered the dollar plate lunch at local restaurants, Beggs frequently returned his slice of pie to the kitchen and demanded a nickel back from the cashier. "A dollar has to work an entire year to earn a nickel," he would remind his companions. Instead of buying his own newspaper, Beggs generally picked up someone else's discarded copy on the streetcar ride to work. Although he lived in Milwaukee for more than 20 years, Beggs never owned a home in the city, preferring to live, at company expense, at the Pfister Hotel. His habitual thrift was not rooted in need. An unusually astute and zealous investor, Beggs owned stock in his own companies and dozens of others that made him a millionaire many times over. But he spent almost nothing on himself and gave even less to charity; generosity was a completely alien concept. "I earned it," Beggs might have said, "and I'm going to keep it."

Beneath the gruff exterior lay a gruff interior. Beggs' only known soft spot was for a group of TM employees he referred to as "my boys," a corps of young men he hired to run errands around company headquarters. When he was in an expansive mood, Beggs would gather them in his office to offer his own life as an illustration of America's possibilities. Retracing his impoverished youth, Beggs would

The field general at the front: John Beggs addressing TM employees at the 1908 company picnic

remind the boys that they, too, could aspire to the splendor that was his as TM's president. "Hitch your wagon to a star" was among his favorite expressions.

This moralizing tightwad did not let his miserly instincts impair his visions of empire. Beggs was, in fact, as grandiose a dreamer as he was a tireless penny-pincher. "Dreamers," he once said, "are essential in this world — if they be working dreamers." His working dream was to make TMER&L the best electric utility in the country, bar none, with a pronounced emphasis on transit service. In his efforts to make that dream a reality, Beggs was absolutely single-minded. He had no seconds-in-command at TM and, although the North American board retained ultimate authority, Beggs was as unstoppable as a force of nature. "Advice, criticism, abuse make little impression on his conduct," wrote the *Milwaukee Sentinel* in 1911. "A stickler for right — as he sees it — is John I. Beggs." Predictably, the historic figure he most admired was Napoleon Bonaparte.

John Beggs was, in short, the ultimate big picture/small picture manager. He had an accountant's mastery of the financial details and an engineer's grasp of the technical details. He possessed a military general's command of both his field and his forces, and a visionary's view of the

Additions to the transit fleet included dozens of up-to-date double-truck cars (left) and the Milwaukee, *a sumptuous private car that Beggs used as a mobile office.*

future in all its glory. Crusty and cantankerous by nature, Beggs practiced an autocratic style that earned him more respect than affection, but he clearly had remarkable abilities.

The man was admirably equipped for the job at hand, but just what did John I. Beggs do? To put it simply, Beggs worked on three fronts simultaneously: He upgraded Milwaukee's transit service, extended TM's reach into the surrounding region, and centralized the company's physical plants. By the time of his first retirement in 1911, Beggs had radically refashioned the system in his own image.

City service was his first priority. Rapid obsolescence plagued the early electrical industry, a situation that has its parallel in the computer field today. Equipment that defined the state of the art one year was practically antiquated the next, and Beggs moved quickly to keep pace with the latest developments in electric transit. Between 1896 and 1900, TMER&L purchased 173 double-truck (eight-wheeled) cars to replace its smaller single-truck models. (Several of the single-truck units were put into service in Racine.) TM's entire fleet soared from 165 cars in 1895 to 450 in 1906. These cars would become icons, objects of veneration, for a later generation of streetcar buffs, but they were simply the best in transportation when they appeared on the streets of Milwaukee.

One addition to the fleet demonstrated that Beggs was entirely willing to spend money — when it was the company's. In 1904 the St. Louis Car Company, then the world's largest streetcar manufacturer, delivered the *Milwaukee,* a showboat that served as Beggs' private car. Outfitted with vermilion and rosewood trim, a bathroom, a full kitchen, and even a fireplace, this palace on wheels was, in the *Street Railway Journal's* considered opinion, "undoubtedly the most costly private car ever built for use on electric railways." Beggs had barely three years to enjoy its regal appointments; the *Milwaukee* was destroyed in a 1907 carbarn fire and never replaced.

As TM's fleet filled out, John Beggs made a minutely detailed study of the local transit system. He patrolled the Milwaukee area in a horse-drawn buggy, noting the growth trends and developing a master plan for his streetcar network. Like the freeway planners of a later generation, Beggs envisioned an "inner belt" ringing the city from Keefe Avenue on the north to Oklahoma Avenue on the south, and an "outer belt" serving the suburban fringe that was sure to materialize around Milwaukee. Electrified belt lines remained a pipe dream, but the system showed substantial progress under Beggs. Heavier tracks were laid, parallel lines were eliminated, new routes were added, and operating schedules were altered. The transit system of the future was beginning to take shape. "This street railway has been to me a matter of pride," Beggs said in 1909, "and one reason why I have stayed with it is to install in this city an ideal system of street car transportation."

Milwaukee's city service grew rapidly under Beggs. New tracks transformed Third Street north of Grand (now Wisconsin) Avenue.

An open-air summer car at 11th and Wells, looking west

The motorman and conductor of Car 196 paused on the Holton Street viaduct in 1906.

TM's tracks converged at Plankinton and Grand (Wisconsin) Avenues. The view is northwest up Plankinton.

Beggs was looking for every possible advantage, and even the smallest decisions were important to him. R.O. Jasperson, his long-time private secretary, recalled Beggs as "a wizard at figures" whose success was rooted in "his insistence upon saving pennies in countless ways until the total mounted to a formidable sum." With larger, faster cars in service, Beggs began to run fewer cars per line at somewhat longer intervals, thereby increasing ridership and reducing the number of trips at the same time. During his first two years at the helm, TMER&L's net income per car-mile increased 156 percent.

Changes in city service were incremental improvements to an existing system. Beggs broke entirely new ground with a considerably larger project: the creation of a full-fledged interurban rail network. Like his counterparts in other cities, he was taking dead aim at the passenger trains of the steam railroads. Milwaukee had had rail service since 1851, but Beggs still saw opportunities almost 50 years later. Offering more frequent service with more frequent stops and substantially lower fares, interurbans could definitely compete with the steam roads, and without a major loss of speed. (Larger, heavier, and more adequately powered than city streetcars, interurban cars could top 60 miles per hour with the controller wide open.) Although Beggs rarely mentioned the TM system's major competitors, his target was clear in advertisements that urged travelers to "Take an Electric Ride — No Smoke, No Cinders, No Dirt."

Purchasing two transit lines from the Pabst brewing family extended TMER&L's reach into suburban Wauwatosa. The view is southwest on today's Harwood Avenue.

His first step on the road to empire was the purchase of TM's last remaining transit competitor in the region: the Milwaukee and Wauwatosa Motor Railway Company. In 1896, when he was still a part-time "expert adviser" to North American, Beggs had urged the purchase in the strongest possible terms, and the deal was closed in January, 1897. The Motor Railway controlled two lines linking Wauwatosa, then a small village two miles west of the city limits, with central Milwaukee. One was an electric line that passed through Washington Park, the other a steam railway (promptly electrified) that entered the city on Wells Street. Both lines were owned by brewer Frederick Pabst, whose summer farm (now the elegant Washington Highlands subdivision) happened to lie in Wauwatosa.

The $770,000 purchase also brought a pair of powerhouses into the TM system: a railway plant in Wauwatosa and a lighting plant on Broadway in downtown Milwaukee. Waste steam from the Broadway plant's boilers was piped underground to provide heat as well as light to the Pabst Theater, the Pabst-owned St. Charles Hotel, and other buildings on the east side of the river. (It was thus the earliest instance of cogeneration — simultaneous production of thermal and electrical energy — in what became the Wisconsin Electric system.) The Broadway plant, after substantial remodeling, also housed TMER&L's general offices until 1906. The Pabst purchase included one more memorable structure: a spindly steel viaduct spanning the Menomonee Valley at Wells Street. TM crews covered its wooden deck with sheets of steel, and the viaduct continued to thrill riders until the demise of streetcar service in 1958.

On December 14, 1896, shortly before the Pabst papers were signed, TMER&L organized a subsidiary to hold its new properties: the Milwaukee Light, Heat & Traction Company (MLH&T). TMER&L and MLH&T were the same company in everything but name, but the division simplified administration; TM concentrated its efforts on the city of Milwaukee, and MLH&T held practically everything outside the city limits.

John Beggs posed in his carriage outside the old Pabst powerhouse in 1902. The building on Broadway served as TM's headquarters for another four years.

The Wells Street viaduct, in regular service from 1892 to 1958, remains one of the most fondly remembered structures in the utility's history.

The next property in Milwaukee Light, Heat & Traction's portfolio was an interurban line to Waukesha, an old resort town west of Milwaukee that was showing new life as an industrial center. Waukesha was clearly an attractive rail destination, but John I. Beggs may have been just as interested in a nearby amusement park, Waukesha Beach. In 1895 a group of Waukesha promoters developed a resort on the south shore of Pewaukee Lake and linked it to their city with five miles of electric railway tracks. Swimming and picnicking were the featured attractions at first, but Waukesha Beach evolved into "The Fun Center of Southern Wisconsin," with a hotel and a dancing pavilion, a bowling alley and a skating rink, arcades and fun houses, and a Ferris wheel that was visible for miles. Streetcar companies across the country were developing similar attractions at the ends of their lines — a trend that failed to move John Beggs. "We do not believe," he told a reporter in 1899, "that it pays street railway companies to run pleasure resorts." Beggs was quick, however, to see the profit potential in a transit line serving someone else's resort. In 1897, after brief negotiations, MLH&T acquired the Waukesha Beach Electric Railway for $62,599.

The Waukesha Beach line would have been practically worthless without a connecting link to Milwaukee, and that link was already under construction. A short-lived competitor was laying track between the two cities in 1897. Beggs bought the line in September and completed its roadbed and overhead to TM standards. Regular service to Waukesha and Waukesha Beach began in June, 1898. On September 25, Beggs led an entourage of Chicago traction officials and journalists on a tour of his new line. The *Street Railway Journal* recorded "the unanimous opinion of all present" that the new railway was "the finest interurban electric line yet constructed." The Waukesha extension was busy immediately, and summer patronage was

The end of the line at Waukesha Beach. The Pewaukee Lake resort was a popular summertime destination for interurban passengers.

almost overwhelming. Scores of Milwaukee firms (including TM) chartered whole trains for their yearly company picnics at Waukesha Beach, invariably flying banners that carried their names and logos. "The electric railway," reported the *Railway Journal* in 1899, "has taken nearly all the business away from the steam railroad," and the local press dubbed the Milwaukee-Waukesha interurban "the gold mine."

Beggs' next foray took the company south. In January, 1896, two weeks before TMER&L rose from the ashes of the Milwaukee Street Railway, a Detroit promoter, Matthew Slush, had joined Racine investors to form the Milwaukee, Racine & Kenosha Electric Railway. The line was completed through Racine and Kenosha in June, 1897, and access to Milwaukee was provided by a connection with the TMER&L system at its terminus in South Milwaukee. Beggs normally resisted ties with other lines, but he had reason to believe that the Milwaukee, Racine & Kenosha would one day be his. That day came in 1899, when persistent financial problems made the line ripe for the picking. On March 1, MLH&T bought the interurban and, with it, the Belle City Electric Railway, Racine's streetcar company, for $400,000. Beggs found fault with the MR&K's "crooked highway" and called the Belle City line "a decrepit, bankrupt property," but both acquisitions were quickly brought up to system standards. By 1900 MLH&T operated on 73 miles of track, reaching south to Kenosha and west to Waukesha Beach.

Beggs concentrated on Milwaukee city service for the next two years, but interurban construction resumed in

The Milwaukee-to-Kenosha line skirted Monument Square in the heart of Racine.

1902. MLH&T headed southwest, serving the only quadrant of Milwaukee that had never enjoyed steam railroad service. By 1903 a new interurban line had reached Hales Corners, then a quiet rural crossroads, and it pushed westward by fits and starts, reaching Muskego in 1904, Big Bend in 1906, and Mukwonago in 1907. Most of the physical labor — spreading ballast, laying track, and setting poles — was done by crews of immigrants, particularly Italians and Hungarians, who were housed in an old Mukwonago sawmill during construction. Their presence in rural Wisconsin led to some interesting cultural exchanges, including one reported by the *Mukwonago Chief* (June 21, 1906):

> *Some of the Italian laborers working on the new electric line here gave an open air concert [at the mill] Sunday night that was very fine. The singing and playing by these men was as good as anything of the kind we have ever heard in this village.*

Two of the bolder players were persuaded to perform on the village square two weeks later, prompting a wistful comment from the *Chief's* editor: "How fine it would be if we too could have a band."

Completion of the southwestern line to its terminus in East Troy was celebrated with an elaborate ceremony on December 13, 1907. John I. Beggs stepped off the inaugural train to the "booming of cannon crackers [and] loud hurrahs and cheers," according to the *East Troy News*. The village cornet band (a point of pride in the rivalry with Mukwonago) led the party to a banquet at the leading local hostelry, where Beggs declared East Troy "the most beautiful small town in the state." The *News* considered the coming of the interurban an unmistakable sign of arrival:

> *Now that we have the railroad, and it is a first-class one in every respect, let us all do our part to advance East Troy, show up her good points, strive to improve in every way possible and thus continue on the upward march so well begun. Tell all the good things about your town you know and show the stranger who enters it that we are a live people.*

Steel rails on gravel roads: an interurban on the southern outskirts of Oconomowoc

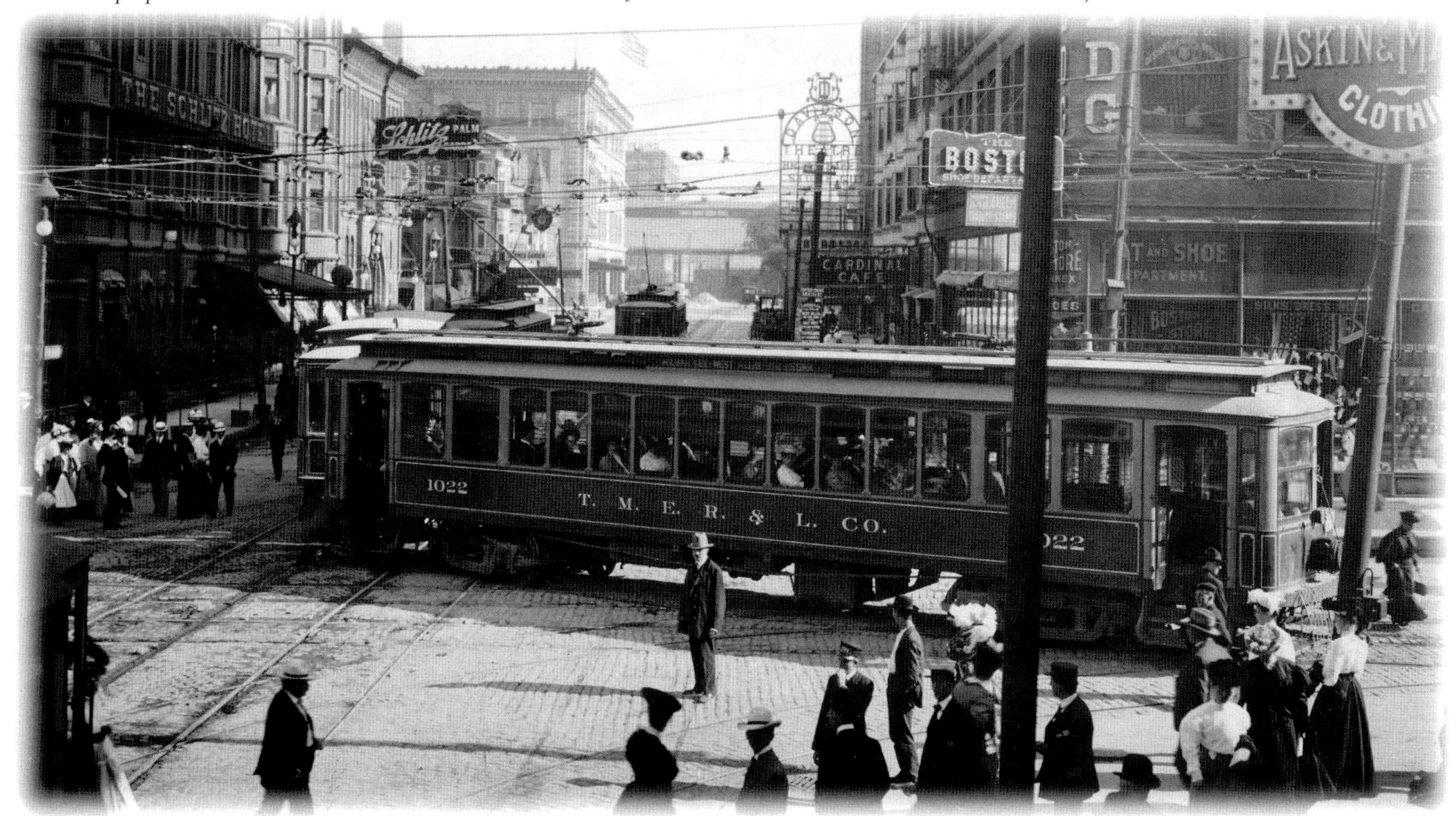

Bound for Waukesha, Car 1022 turned onto Grand (Wisconsin) Avenue from Third Street in 1906. The view is south on Third.

Cutting and filling for the Mukwonago line in 1906 required steam locomotives and a small army of workers, many of them Italian immigrants.

Hail the conquering trolley: Watertown residents greet their first interurban car on July 30, 1908.

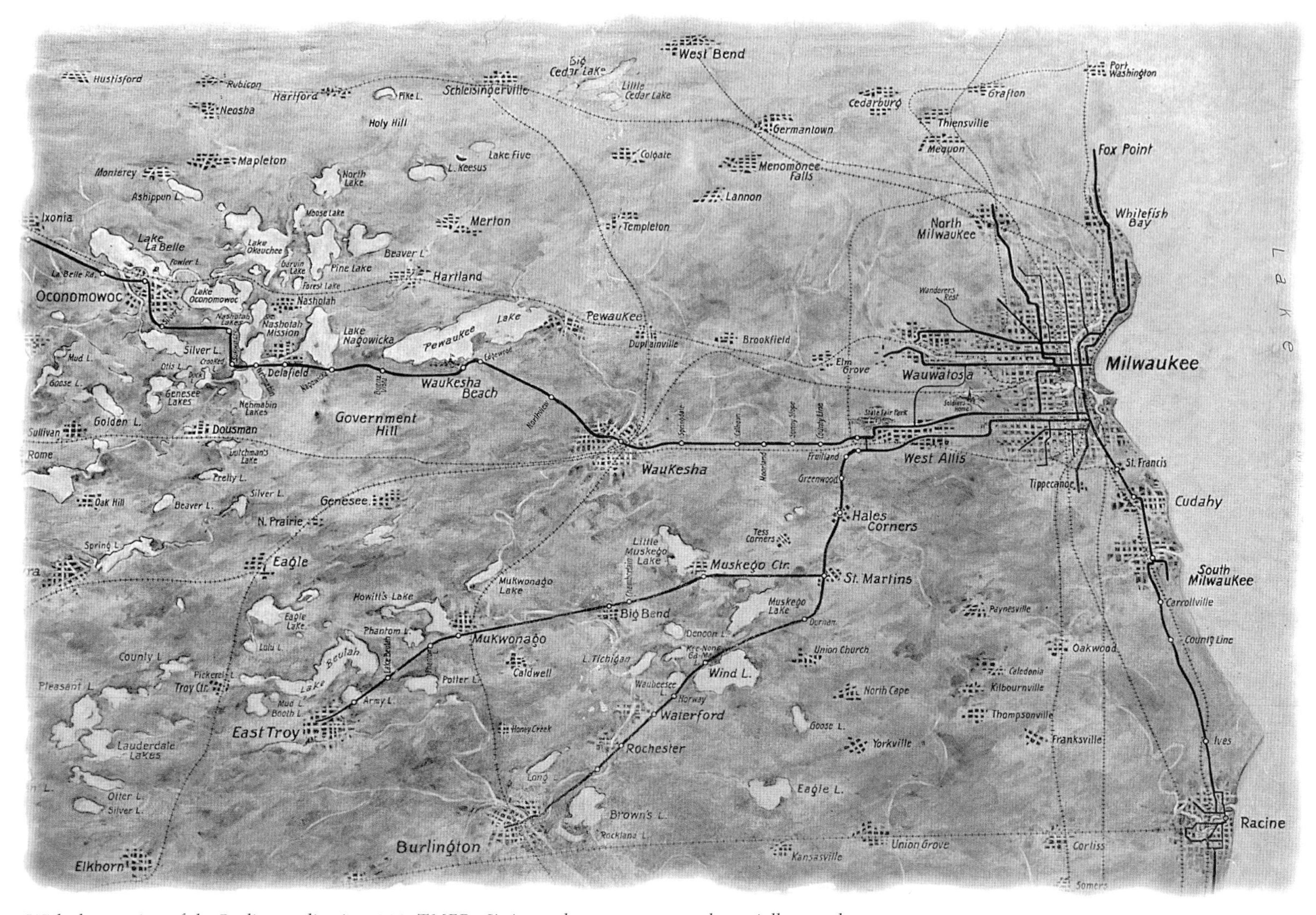

With the opening of the Burlington line in 1909, TMER&L's interurban system was substantially complete.

MLH&T enlivened other towns in Milwaukee's hinterland during the same period. The Waukesha Beach interurban edged steadily westward across the Kettle Moraine, following a circuitous route over hills, between lakes, and around farms. The line reached Oconomowoc in 1907 and its terminus in Watertown in 1908 — more than halfway to Madison. The Watertown trains traversed the lake country of the Hartland-Oconomowoc area, and the *Street Railway Journal* (Aug. 13, 1907) predicted an increase in vacation traffic: "The new lines ... will afford Milwaukee and Chicago people a convenient means of reaching some of the most attractive spots of a most picturesque and romantic region."

Beggs turned southwest again for his final extension, a branch line to Burlington. Joining the East Troy interurban at the little town of St. Martins, the Burlington branch was completed in 1909. The system would grow by accretion in later years, but MLH&T reached a peak of 232 miles of track in 1910 — a threefold increase in just 10 years. The Midwest was the nation's center of interurban railway development and, thanks to John I. Beggs, Milwaukee had one of the finest systems in the Midwest.

The interurban lines helped to close the historic gap between city and country. From Wind Lake on the south to Lac La Belle on the west, they opened the countryside to weary city-dwellers seeking the balm of blue waters and open skies, and they made the big-city resources of Milwaukee available to thousands of rural residents. The East Troy interurban offered a late car from downtown Milwaukee (leaving at 11:15 PM) "to accommodate theater-going patrons." (The two-hour ride cost 70 cents, or $1.25 per round trip.) With the advent of freight service, farmers and artisans looked increasingly to the Milwaukee market. A Big Bend cheese factory was forced to close when local dairymen decided to ship their milk, via the interurban, to a Milwaukee dairy that paid substantially more per hundredweight, and East Troy developed a modest industrial base directly dependent on MLH&T lines.

Power and light generally followed the trains; the interurbans were important agents of rural electrification. On July 1, 1911, the streetlights went on in East Troy. "EAST TROY BECOMES CITIFIED," blared the banner headline in the *News*, and the accompanying story described the "big jollification" held to mark the event. Nearly 3,000 people gathered in the village square to watch the lights come on, with additional entertainment provided by TMER&L's 42-piece employee band. The local opera house was soon offering an "electric theatre" twice a week, showing silent films shipped from Milwaukee, and Griste's Drug Store unveiled an electric freezer that could turn out 40 gallons of ice cream a day.

Although the interurbans changed thousands of lives, both urban and rural, the system must be viewed in context. Milwaukee operations remained the core of Beggs' empire; in 1909, when the regional network was substantially complete, TMER&L generated nearly three times the revenue of its MLH&T subsidiary. Nor did the interurban system ever achieve the scale that John Beggs had planned for it. Early in his tenure as general manager, Beggs had developed what he called a "mental map" of regional transit lines. He envisioned Milwaukee as the hub of a gigantic wheel, with interurban spokes radiating outward as far as Chicago, Lake Geneva, Madison, Fond du Lac, Sheboygan, and Green Bay. Few of those spokes were more than half-finished by the time Beggs left office, the result of competition in some cases, a shortage of capital in others and, in the end, a power shift on the North American board.

Officials of the Appleton Edison Company showed off the new trestle at the eastern end of their streetcar line. The entire company came under John Beggs' control in 1900. (Outagamie County Historical Society)

It was not for lack of trying. Although he generally relied on the North American Company to fund his dreams, Beggs was entirely capable of acting on his own. In 1900, for instance, he headed a private syndicate that purchased the lighting and traction utilities in Appleton. The Fox River city had already distinguished itself as a pioneer in both hydroelectric power and electrified transit, but local firms had shown a greater talent for making history than for making money. Beggs shored up local service and made Appleton the hub of a short-line interurban extending from Kaukauna to Neenah. His clear intent was to link Appleton with the MLH&T network in southeastern Wisconsin — Beggs went so far as to publish a map outlining his proposed route — but Appleton remained the isolated outpost of an empire that never matched its emperor's dreams.

John Beggs did not take over the world, or even Wisconsin, but his aggressive expansion program had a formative impact on the Wisconsin

Electric system of the present. Its lasting significance is geographic. In the vast majority of communities the interurban touched, from Racine to Watertown to Appleton, Beggs offered power and light as well as transportation, either by inaugurating local service or by purchasing companies already in business. Traction opened the door to everything else; the interurban network was, in effect, the skeleton of a full-service regional utility. Long after the heyday of steel-wheeled transit, Wisconsin Electric continues to serve a geographic area stretched out over the framework of interurban lines built by John I. Beggs nearly a century ago.

The master's third great task — centralizing TM's physical facilities — followed naturally from the first two. As Milwaukee service grew and the interurban system expanded, Beggs had a desperate need for more generating capacity, more people, and more room. Unplanned obsolescence was a continuing headache. When the Edison powerhouse on the river commenced service in 1890, it was one of the finest in the country, but its generators were practically museum pieces only a decade later. The world, for one thing, was moving rapidly from direct current to alternating current. (Pioneered commercially by Thomas Edison, direct current flows continuously from one pole to the other, and early systems were limited by their wiring to a radius of perhaps two miles. Alternating current, popularized by George Westinghouse, changes direction constantly between its poles, and it can be stepped up or down to provide power many miles from its source.) Nonexistent in 1892, AC systems produced just over 50 percent of Wisconsin's electrical power in 1902 and 71 percent in 1912.

John Beggs actively abetted the trend. In 1898 TMER&L began construction of a new powerhouse on the Milwaukee River at Oneida (East Wells) Street, adjacent to the 1890 plant. It featured the latest in DC generators and, in 1903, a storage battery that occupied an entire floor of the building. (The battery room was lined with acid-proof cement as a precaution.) Soon, however, a move to AC was afoot. In 1903 the company opened a new powerhouse upriver at Commerce Street, squarely on the site of the pioneer Badger and Hinsey plants. The Commerce Street facility was described as a "transitional" powerhouse. Direct current was produced in its north half and alternating current in its south half, and the generators were driven by both old-fashioned steam engines and newfangled steam turbines. In a series of three expansions between 1905 and 1912, the balance of power in the plant shifted steadily from DC to AC.

In 1900 the old Edison powerhouse on the Milwaukee River was joined by a new plant at Oneida (East Wells) Street. The building is still in use as the Milwaukee Repertory Theater's home.

(Milwaukee Public Library)

Over the same period, pistons gave way to propellers as steam engines were replaced by turbines. A decade after it opened, the Commerce Street powerhouse had a capacity of 62,000 kilowatts — 15 times TM's entire capacity in 1896 — and it had become the workhorse for the whole system.

A smaller AC plant — the old Pabst powerhouse on Broadway — had been part of the system since its purchase in 1897. In a board meeting held just before the deal was closed, Henry Payne had demonstrated why TMER&L needed someone like John Beggs. Exposing the limits of his technical competence, Payne suggested that TM use the Broadway plant for incandescent lighting and the riverside plant for traction. Beggs noted quietly that the lighting system ran on direct current and the Broadway facility produced alternating current. Payne's proposal was therefore, he said, "neither economical nor practicable."

The Broadway plant became, among other things, the firm's corporate office, but it was far too small to accommodate a company growing as fast as TMER&L. The total number of employees soared from roughly 1,200 in 1896 to more than 3,000 in 1911. Most were motormen and conductors but, as the management staff swelled, the Broadway office began to feel like a sardine can. Beggs had decided early in his tenure that TMER&L needed a genuine headquarters, a place that would function as the nerve center of the entire system. In January, 1897, months before he became general manager, Beggs had urged the North American board to build "a central car station in which all railway lines will concentrate." By 1899 he had found his site: a square block between Second and Third south of Michigan Street. The only building on the parcel was St. Gall's Church, the home of an old Irish congregation that had merged with Holy Name, another Celtic stronghold, to form Gesu Parish in 1895. The church was surplus property; TMER&L bought the block and began to make plans for a new Milwaukee landmark. The site was, interestingly, just across the street from the depot of the Milwaukee Road, whose local passengers were defecting to electrified transit in droves. When TM's building was finished, Beggs could look out his window and tell the time from the railroad depot's clock tower.

John Beggs envisioned the ultimate multi-purpose building on the St. Gall's site. In 1901, when planning began in earnest, it was described as a "central car station and office building." In 1902, when foundation work

TMER&L built a second new power plant around an old "porcupine" boiler (left) from the Badger Illuminating facility. Completed in 1903, the Commerce Street plant quickly became the system's workhorse.

commenced, TM officials were calling it "the new general office building, central car house and terminal station." The structure was a good deal more than that. Working closely with architect Herman J. Esser, Beggs created a brick-and-stone monument that embodied his own conception of TMER&L's functions and importance. The basement housed a cogeneration plant that provided electricity and steam heat for TM's headquarters and other downtown buildings. (The East Wells plant shared the steam load.) The ground floor served as the depot for the entire interurban system; an elegant waiting room, lined from floor to ceiling with Tennessee marble, opened onto a series of 11 tracks running completely through the building. The second floor featured facilities for the entertainment and edification of TM's employees: a 1,200-seat auditorium, bowling alleys, dining rooms, a library, billiard rooms, lockers, and even a barber shop. The third floor housed everything from employee medical facilities to a machine shop that produced electrical switchboards. The fourth floor, the pinnacle of the building, was John Beggs' domain. His corner office and the adjoining board room were as opulently appointed as the city's finest mansions, with hand-carved moldings, parquet floors, and massive fireplaces.

The Public Service Building, as it was called from the beginning, was ready for occupancy in 1906. Beggs considered the neoclassical landmark his crowning achievement, a blend of form and function that centralized his far-flung empire, boosted employee productivity, and heightened TM's profile in the community. One architectural detail, a stained-glass window above the main entrance, symbolized his view of the building; it depicted a swarm of bees buzzing around an oversized hive. The Public Service Building was the beehive for a busy enterprise and, during John I. Beggs' reign, no one in the swarm had to ask who played the role of the queen bee.

John Beggs chose the site of St. Gall's Church, an old Irish parish at Second and Michigan Streets, for TMER&L's new headquarters. The Milwaukee Road depot (left) was just across the street. (Milwaukee Public Library)

From site work through steel work, construction of the Public Service Building took nearly three years.

Completed in 1906, the PSB became an instant Milwaukee landmark.

Work spaces in the new building ranged from the high ceilings of the Ticket Auditing Department ...

... to the opulent confines of John Beggs' private office.

The Public Service Building's focal point was an elegant auditorium, shown here just before a formal banquet. An electric train made the rounds of the table during dinner, and each place setting featured an electric plate warmer. Even the miniature streetlights worked.

The new building attracted its share of attention. The *Street Railway Journal* (July 14, 1906) had nothing but praise: "It is noteworthy, even in this age of magnificent buildings, for the attractiveness and taste of its architecture, the beauty of its interior decoration, and the thoroughness of all its appointments." *Architectural Record* (April, 1908) gave the PSB national standing: "Under its roof are carried on a greater variety of occupations than in any other building in the country."

A stained-glass window above the entrance signified John Beggs' view of the new building.

The Public Service Building was a train station, a powerhouse, an employee social center, a machine shop, a theater, a medical clinic, and a corporate headquarters rolled into one, but John Beggs planned even greater things. One persistent legend has it that he designed the building for easy conversion to a hotel — just in case the electricity business didn't work out. Beggs had no such doubts about the future of power. What he envisioned was a multi-use building that would provide a home for both his own company and for income-producing tenants — much like the great train stations of New York City. Until TM needed the space, rooms on the upper floors were leased to outside businesses, including a jewelry store, and the auditorium was regularly

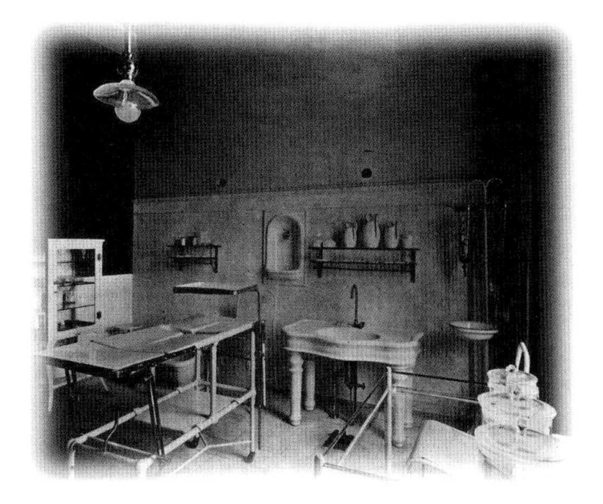

The PSB also contained facilities for the health and welfare of TMER&L's employees, including (top to bottom) a well-equipped medical clinic, a billiard room and, in the best Milwaukee tradition, a bowling alley.

rented for conventions and community events. Beggs wanted more of the same. Looking ahead, he had put enough steel in the four-story structure to support eight additional floors. Beggs made his plans crystal-clear when he testified in a 1909 legal proceeding. The *Electric Railway Journal* (March 6, 1909) paraphrased his direct comments:

> *Inside of ten years the building would carry itself and pay a dividend to the company without a loss to it. Only one-third of the building had been constructed and with the upper stories added the building would make the best hotel, apartment building or office building in the city.*

Like the thrifty Germans who were lining Milwaukee's streets with duplexes, Beggs wanted his tenants to help with the mortgage payment. His expansion plans came to nothing but, 90 years after its construction, the PSB continues to buzz with activity as the administrative beehive for the Wisconsin Electric system.

The downtown edifice was a highly visible, highly personal expression of Beggs' penchant for control, efficiency, and prestige. His last major building project — the Cold Spring car shops — stressed pure function. Since 1894 the system's repair and service facility had been a shop on Kinnickinnic Avenue, near the site of the old Cream City horsecar barns. As TM's explosive growth led to hopeless overcrowding, Beggs began to look around the city for room to expand. He found a centrally located site, a former brickyard at 40th and Cold Spring (now McKinley Boulevard), and directed TM's engineers to design "a complete shop second to none in the country." Construction began in 1911, and the complex was finished a year later. The Cold Spring shops contained everything necessary to service a fleet that had grown to 630 cars: a blacksmith shop, a machine shop, repair pits, transfer tables, erecting bays, and a paint shop. With 300,000 square feet under roof, the plant was nearly as large as two nearby giants: the Miller brewery and the Harley-Davidson motorcycle factory. The tradesmen of the Cold Spring shops handled routine maintenance and repair work, but they also had the facilities to refit old cars and build new ones from scratch. Beggs, typically, wanted to control his own source of supply.

Completed in 1912, the Cold Spring shops had ample facilities for the maintenance, repair, and reconstruction of TM's transit fleet.

The three central building projects of Beggs' presidency — the Cold Spring shops, the Commerce Street powerhouse, and the Public Service Building — underscored a dominant theme of his administration: an unremitting emphasis on traction. Cold Spring made and repaired the cars, Commerce Street powered them, and the Public Service Building functioned most visibly as an interurban rail depot. Lighting and power were distinctly secondary. Beggs had his priorities carved in stone on two medallions above the PSB entrance; one depicted a horsecar from 1890 and the other a thoroughly modern streetcar from 1905. Under his leadership, "Electric Railway" was TMER&L's middle name.

Given his earlier success as an electric lighting executive, Beggs' accent on transportation was surprising. He did not, however, completely ignore other markets for electrical power. TM had a lucrative street-lighting contract with the City of Milwaukee, and the Public Service Building housed a retail store that offered an assortment of lighting fixtures, irons, fans, vacuum cleaners, and other appliances. The company also provided power for at least one spectacular display of lighting's potential. Foreshadowing later events like Summerfest, Milwaukee staged an elaborate mid-summer carnival in 1900. Downtown streets and buildings, reported the *Milwaukee Sentinel* (June 27, 1900), were lit up like a Christmas tree, inspiring wonder in the gathered throngs:

> *Men who have survived the age when the tallow dip was the great illumination for the people marveled ... as they looked up and saw the great buildings set out in lines of fire, the streets canopied with lines and arches of electric light, and business emporiums festooned in red and white and green and orange colored shining globes.*

Two stone medallions above the PSB's front door made John Beggs' priorities crystal-clear.

Although Beggs emphasized the transit side, TMER&L also sold a full range of power-hungry appliances. These young women were putting on an "electric cooking demonstration" at the Public Service Building in 1907.

One of the carnival's highlights was an "Electrical Pageant" that drew 200,000 spectators. A night-time parade of 20 floats, powered and lighted by TM's trolley lines, wound its way through the city, bearing likenesses of the battleship *Wisconsin*, the Goddess of Light, Mother Goose characters, and even Jonah and the whale — all outlined in incandescent lights. Nearly 6,000 bulbs were put to use, including 900 for the whale alone. The pageant, declared the *Sentinel*, "records an epoch in the history of papier-mache and illuminative possibilities."

Its possibilities for pageantry aside, electricity's hold on the population was not yet complete. Only one American home in 10 had electrical service in 1910, and a TM ad of the time underlined its appeal to the affluent: "Have an Electric Home. This Solves the Domestic Help Problem." Industrial users offered another avenue to sales growth, one that TM failed to take initially. Milwaukee's factories and foundries electrified rapidly after 1900, but John Beggs was quite content to let them build free-standing power plants of their own. Much later than most of his counterparts in the industry, Beggs persisted in the belief that the core business of an electric utility was electric transit. In 1902 the "railway department" generated fully 83 percent of the TM system's gross revenue; in 1909, despite a surge in demand for non-transit power, the proportion had risen to 85 percent.

Beggs might be criticized, in hindsight, for his insistence on taking the trolley to profits and prosperity, but no one could argue with his achievements. He took a system still emerging from the depression of the mid-1890s and turned it into an expertly managed, dividend-paying transportation empire. Between 1896 and 1911, the Milwaukee system grew from 130 miles of track, nearly all within the city limits, to 376 miles stretched across eastern Wisconsin. The number of passengers carried rose from 28 million a year to 132 million. System-wide employment jumped from perhaps 1,200 to 3,125. Beggs' expansion campaign cost North American $23 million (roughly $400 million in 1996 dollars), but its results were apparent on the bottom line. Gross revenues soared from $1.47 million in

The Butterfly Theater, a landmark on Grand (Wisconsin) Avenue at Second Street from 1911 to 1930, offered a spectacular example of electric lighting's possibilities.

1896 to $6.65 million in 1910, and net income jumped from $459,000 at the beginning of Beggs' tenure to a peak of $1.29 million in 1909.

Impressive as they are, the numbers do not begin to demonstrate the TMER&L system's impact on the region it served. Superlative streetcar service broadened the average Milwaukeean's choices of residence, employment, and recreation, eroding the sectional rivalries that had marked the city since its pioneer period. The cars also hastened the decentralization of an unusually dense community; for suburbs from West Allis and Wauwatosa to North Milwaukee and South Milwaukee, trolley lines were literally lifelines. In the surrounding region, interurban service helped to break down the barriers between urbanites and rural residents, opening, for both, avenues to a larger world.

No one was more aware of the system's impact than John I. Beggs himself. On the eve of his first retirement in 1911, Milwaukee's Napoleon looked back with pleasure. "It has been one continual, triumphant march," he declared, and he laid credit for Milwaukee's advancement squarely at his own doorstep. "Our interurban lines have made Milwaukee," Beggs told a reporter, "and they have served as an object lesson in other sections of the country.... Thinking people know that the Greater Milwaukee is due to the extension and development of our system." Broadly speaking, he was right.

Every self-respecting commercial district used electric lights to attract customers. This view shows Third Street near Grand Avenue in 1906.

II. Crusaders And Infidels

The other side of the coin was endless public turmoil. TMER&L may have been a Dr. Jekyll to its investors and supporters — a prosperous, progressive railway system — but it was Mr. Hyde to a much larger number of detractors. For all its obvious achievements, the company was widely viewed as an incarnation of corporate evil. Detractors loved to paint Henry Payne, John Beggs, and all their associates as infidels, and a shifting cast of self-righteous crusaders mounted a holy war against the system's leaders.

The crusade had its roots in the Panic of 1893, years before Beggs' arrival. Milwaukee had become a major industrial center in the quarter-century preceding the depression and, as long as the city prospered, only diehard idealists questioned the old rules of laissez-faire capitalism and political profligacy. But industrialization and its twin, immigration, had changed the context in which those rules applied. When the economy collapsed, civic leaders were not prepared for the consequences: unemployment approaching 40 percent, an overburdened relief effort, and general unrest and suffering. Milwaukee was a city in crisis, and the social chaos seemed to indicate that something was wrong, radically wrong, with the whole system.

Answers were soon forthcoming. As disenchantment with the status quo swept the community, the lonely cries of the idealists became a general chorus, and support grew for a structural reformation of Milwaukee's public life. There was general agreement that the city faced three primary evils: concentration of political power in the hands of corrupt public officials; concentration of economic power in the clutches of self-serving corporations; and a pervasive, unholy alliance between the two groups. A broad-based coalition of labor leaders, Republican insurgents, newspaper editors, civic reformers, and socialists mounted a campaign to curb the power of the corporations and reclaim the levers of political control for the people.

The Milwaukee Street Railway (and TMER&L, its direct descendant) served as a ready-made villain in the morality play that followed. Like its counterparts in other American cities, the streetcar company was the enterprise that citizens loved to hate. It was a monopoly, first of all, and, to make matters worse, it was owned by eastern capitalists. (Milwaukee's Teutonic citizens might have rested easier had they known how much of the firm's debt was held by German banks — a legacy of Henry Villard's connections.) But the company's Achilles' heel was its status as a bastion of the old order. The Street Railway was a product of the Gilded Age, a time when business and politics were so intertwined that they were practically indistinguishable. Henry Payne was chosen to head the system because of his political skills, not his technical competence. As the tide of anti-corporate sentiment continued to rise, Payne and his company found themselves transformed into potent symbols of all that was considered wrong with the American system.

National reformers had other corporate targets, of course — the sugar trust, the tobacco trust, the railroad trusts — but those conglomerates were little more than abstractions to Milwaukee voters. The Street Railway was the intimately local representative of corporate America. No other private company had a fraction of its impact on the community, and no other private company was so thoroughly enmeshed in local politics. A few salient points were conveniently overlooked: the material risks taken by the system's creators, the dramatic improvements in transit service, and the severity of MSR's financial problems. None of these facts stirred the least bit of sympathy in the system's critics; practically by circumstance, the company became the specific scapegoat for the sins of capitalism in general.

Henry Payne remained true to his colors throughout the turmoil. He was a political operative at heart: a gatherer of votes, a molder of coalitions, a purveyor of influence. Either unwilling or unable to admit that the climate had changed, Payne continued to play by the old rules. His actions on behalf of the company only enhanced its role as a lightning rod for the strikes of reformers. Payne's personal reputation began to sink in the depths of the depression, when he took some highly unpopular steps to save the Street Railway from bankruptcy. Workers were laid off. Tracks on the least profitable city routes were pulled up. Books of 25 tickets for one dollar (the equivalent of a four-cent fare) were pulled off the market. Street-lighting rates were raised. Planned route extensions were canceled, and weeds began to sprout in subdivisions laid out in hopes of streetcar service.

Payne was simply doing what he considered necessary to save the system, but Milwaukeeans came to view his actions as a blatant expression of corporate arrogance. Reformers sensed an opening. City officials, facing a revenue shortage of their own, more than doubled the Street Railway's tax assessment in 1894, and they even courted, for a time, a competing transit operator from outside the city. The reassessment was voided by the state Supreme Court, and the threat of competition never materialized. The battle, however, had just begun, and Henry Payne continued to provide the enemy with ammunition. In 1895 he helped to ensure passage of a state law exempting street railways from local property taxes. (He reported to North American that the measure was "very advantageous to the interests of the Company.") Largely viewed as tax evasion rather than an exemption, the

Accustomed to back-room deals and private "understandings," Henry Clay Payne found the highly public controversies of the 1890s most unpleasant.

measure did nothing for Payne's popularity in Milwaukee. His next step was to secure a two-year delay in the enforcement of a state law that required the Street Railway to enclose the motorman's vestibule in all its cars. A fatal accident had prompted the law. On a sub-zero day in February, 1895, a Milwaukee motorman, half-frozen in his open vestibule, had plunged his car into the Kinnickinnic River, killing himself and two passengers. Memories of the tragedy were still fresh, and Payne's opposition to the vestibule law was viewed as, at best, insensitive. His ham-handed approach to public relations only added fuel to the fire that reformers were trying to light under his company. Payne himself, wrote historian David Thelen, was "the most hated man in Wisconsin at the end of the century."

The 1896 organization of TMER&L as the Street Railway's successor changed nothing at all. Payne, in fact, found himself dealing with rebellion from an unexpected quarter: his own employees. Emboldened by the return of prosperity and the shift in public sentiment, TM's motormen and conductors demanded a raise in pay (from 19 to 20 cents an hour) and recognition of their union, the Amalgamated Association of Street Railway Employees. The demands were delivered on May 1, 1896. Three days later, when Payne showed absolutely no interest in negotiating, nearly all of TM's 750 trainmen went out on strike, and all 300 powerhouse workers soon followed. The struggle that followed was one of the most memorable in the wave of streetcar strikes that swept America's cities at the turn of the century.

Payne's willingness to let his motormen freeze in their open vestibules was widely viewed as the cause of a fatal streetcar accident in 1895. (Milwaukee County Historical Society)

It was clear from the beginning that management would not bend. North American's officers dismissed the strikers' demands as "unreasonable, unfair, impracticable and impossible of being acceded to by the Company." When Henry Payne was urged to compromise with the union, he huffed, "Arbitrate? The company has nothing to arbitrate." Sensing trouble, he had arranged to import streetcar men from Chicago, Minneapolis, Grand Rapids, and other Midwestern cities as strikebreakers. They began to pour into Milwaukee, nearly 1,200 strong, on the day after the strike was called, to a decidedly chilly reception. The union, Payne discovered, had overwhelming popular support. When local hostelries refused to house the new men, he was forced to convert TM's seven carbarns into dormitories, served by a crew of 50 black cooks. When local grocers refused to sell the company provisions, Payne had to buy supplies in Chicago. When the strikebreakers took their cars out onto the streets, they generally encountered barricades, cut trolley lines, spiked switches, and angry mobs. Davy Jones, a Welshman with service in the British army, told a reporter that it was like entering a war zone:

> *We have been through the fire now for two days. I got all my windows broken both days. Bricks and stones rained like hailstones around our heads. Their cries are*

deafening. We are called all the names, curses and swears you ever heard in your life, but we care not. I am here and shall hold my motor or die.

Faced with stubborn replacements and an intractable management, the union adopted a more creative strategy: a general boycott. The strikers urged regular TM passengers to stay off the cars, and the public response was spectacular. Milwaukeeans began to sport badges declaring "To Ride Gives Me a Payne" and "I'll Walk; Will You?" Druggists refused to sell soda water or cigars to anyone seen on a streetcar. A dry goods merchant suspected of treason offered a $100 reward to anyone who could prove "that I have been seen riding on the cars ... since the strike was declared." When Mayor William Rauschenberger urged a conciliatory stance, his butcher refused to sell him meat. The strikers gave their supporters a novel transit alternative. They rounded up every spare omnibus, express wagon, and dray in the surrounding region and began to offer regular passenger service on TM's regular routes, at the same nickel fare. Streetcars whizzed by practically empty.

The strike turned into a two-month war of attrition. Already on shaky financial ground, TMER&L was losing money by the basketful, but the system's leaders refused to negotiate. North American executives, including Charles Wetmore and William Nelson Cromwell, spent weeks in Milwaukee, and they had nothing but praise for "the fidelity, zeal and efficiency" demonstrated by Henry Payne "throughout all the trying circumstances." The strikers, who regularly hanged Payne in effigy, had lost their livelihoods, but they were just as obstinate as TM. When the company offered to recall a third of its men at the old rates, the union adamantly refused.

What finally ended the strike was a combination of fatigue and violence.

During the bitter transit strike of 1896, union sympathizers blocked the streetcar tracks with disabled wagons several times a day. A crowd gathered at Water and Wisconsin Streets to watch the commotion.

After the stalemate had dragged on for a month, some rifle-toting revolutionaries took matters into their own hands. They ambushed a Cudahy car on June 3, seriously wounding both the motorman and the conductor, and fired a day later on a Walnut Street car, injuring a passenger. Public sympathy eroded rapidly, and so did the solidarity of the strikers. Their ranks thinned by the day and finally evaporated. With a substantially new work force in place, the company continued to operate; ridership was back to normal levels by July.

Breaking the union did not endear Henry Payne and his associates to the people of Milwaukee. The conflict continued, but its focus shifted to legislative chambers and courtrooms. City officials and civic leaders, including a number of businessmen, pursued any and all means available to break the company, to bring it to heel. Municipal ownership, their long-range goal, was not a realistic option; the City of Milwaukee, still recovering from a major depression, lacked the resources to either buy TMER&L or build a system of its own. Competitors were actively encouraged to enter the market, but no newcomer could hope to prevail against a company as entrenched as TM. With those major strategic options closed, local officials engaged the company in guerrilla warfare. In 1896 and 1897, the Common Council passed a series of measures designed to harass and embarrass TMER&L. One ordinance required the company to drop its fare from a nickel to four cents. (TM promptly contested the ordinance in court.) Another mandated "owl cars" into the wee hours of the morning. Yet another attempted to repeal franchises on streets not yet served by the company. City officials also supported efforts to repeal TM's 1895 property tax exemption. That attempt failed, but the 1897 state legislature did pass a 4-percent tax on gross receipts over $750,000 — the same standard applied to steam railroads.

Mayor David Rose, a political opportunist who played a key role in the streetcar disputes of the 1890s

The company did what it could to resist the onslaught. On April 17, 1896, the North American Company board resolved that "we should stand absolutely firm in resisting any ordinance tending to reduce fares or to authorize competition or oppress us in any other way." It was the four-cent fare ordinance that corporate officials took most seriously. In a city dominated by workingmen, streetcar fare was not small change. The average worker earned between one and two dollars a day (for 10 to 12 hours of work), and a nickel in the 1890s would buy a good-sized bath towel, an embroidered handkerchief, or a yard of striped sateen. A 20-percent cut in transportation costs had enormous appeal. The company protested loudly that it could not afford the corresponding decrease in revenue. TMER&L was indeed operating on a shoestring; its stock was trading for a fraction of its book value, and holders of common shares had to wait until 1902 for their first dividends. But Henry Payne's pleas of poverty rang hollow to most Milwaukeeans. They had little faith in the figures he presented — sometimes with good reason. Payne told reporters that TM's net income for 1898 was "but $38,108.67," when the company's books, closed to the public, showed a figure of nearly $459,000.

As the four-cent case wound its way through the courts, TMER&L became a major factor in the 1898 mayoral election. Faced with an open revolt in his own party, Payne saw to it that a non-controversial candidate, tinware manufacturer William Geuder, carried the Republican standard. The Democrats chose David Rose, a flamboyant attorney who made TMER&L his real opponent. Rose charged that Geuder was the candidate of the "Republican machine," a machine "made up of the moving spirits of the streetcar ring" — a direct reference to Henry Payne and his fellow TM director, Charles Pfister. Electing his opponent, said Rose, "would be equivalent to the election of the officers of the streetcar company to the office of mayor." David Rose won by a landslide, and harassment of the company continued. In May, 1898, when the U.S. Circuit Court voided the four-cent ordinance as "unreasonable" and "confiscatory," Milwaukee's Common Council promptly repassed it.

The North American Company had had enough. Still operating under an older view of the world, the holding company's officers had long been mystified by the unreasoning antagonism toward their Milwaukee enterprise. When it became clear that the constant turmoil was scaring investors away, President Charles Wetmore sued for peace. In June, 1898, he began to negotiate directly with Mayor Rose for "a comprehensive settlement" and "a cessation of hostilities." "We are unable to understand," wrote Wetmore, "why the municipality should not deal with the most important undertaking within its limits as a man of good business sense would do under similar circumstances." What North American wanted was a definite, long-term foundation for its relations with the city, and to that end Wetmore put a proposal on the table late in the year. It called for an extension of all existing franchises through 1924, annual payments of $100,000 to the City after 1903, and a one-third share in TM's profits after payment of a 6-percent stock dividend. David Rose promptly endorsed the package.

It was a good deal, but Wetmore's proposal lacked one crucial component: a four-cent fare. Reduced fares had become a Holy Grail for the reform coalition, and its leaders protested loudly that Rose had sold out the city. Countering that he had bargained in good faith, the mayor published his complete correspondence with Wetmore in the local papers, to no effect. As the *Street Railway Journal* (December 15, 1898) reported, "So great an outcry was made against this agreement by the press and public generally that the company withdrew its proposition."

Rose and Wetmore went back to the bargaining table. In July, 1899, North American submitted a new proposal: an extension of existing franchises through 1934 (10 years longer than first asked), franchises for 12 new routes, and a phased-in system of four-cent fares. TMER&L would offer 25 commutation tickets for one dollar during rush hours until 1905, when the four-cent tickets would become universal. (Milwaukee's rush hours in 1899, a more toilsome time than our own, were 5:30 to 7 AM and 5 to 6:30 PM.)

The four-cent offer was designed to appease TM's opponents, but the proposal met unexpected criticism. The sticking point was the extension of all franchises through 1934. John Butler, Milwaukee's leading voice of civic reform, dismissed the deal Rose had cut as "the merest mess of pottage." The Federated Trades Council, Milwaukee's leading labor organization, blasted it as one more of TM's "grabbing monopoly schemes ... to rob the poor and fatten the Eastern and local bondholders of the company." In some quarters, the company could do no right. The reformers started a war of injunctions, but the franchise ordinance, with crucial help from Henry Payne and his operatives, passed on January 2, 1900. John Butler lamented, "Thus the introduction of public ownership has been delayed for 35 years, the present company is established in its monopoly during that period, and the public right to regulate fares has been surrendered."

The franchise controversy gave Henry Payne a dubious kind of immortality. Novelist Charles Lush used Payne as the model for his villain, streetcar magnate Henry Bidwell, in *The Autocrats*, published by Doubleday in 1901. The novel has more plot twists than a soap opera, but it features Bidwell as the leader of a sinister cabal bent on passing a long-term transit franchise. With the active assistance of Mayor David "Thorn," the veteran string-puller succeeds, and the result is an ordinance that one of the more virtuous characters blasts as "a piece of vicious legislation [that] takes away from the city by force of contract the rights of citizens yet unborn."

No one seemed to notice, or care, that the company had effectively frozen its revenues for 35 years. But TM officials, notably John Beggs, had done some figuring before the proposal was submitted. Beggs estimated that 60 percent of the system's passengers would use the four-cent tickets, which, offset by an expected increase in volume, would result in an annual revenue loss of only $18,592. He did not figure on the average Milwaukeean's inborn frugality. When the tickets were offered for sale, 81 percent of TM's customers used them exclusively. It took a dramatic increase in ridership — one of Beggs' singular achievements — to make up the shortfall in revenue.

With passage of the 1900 ordinance, there was a lull in the action. Although his reform-oriented supporters defected en masse, David Rose narrowly won re-election as mayor. (One of his campaign promotions in 1900 was a streetcar ticket-holder stamped with a rose and the motto, "A Penny Saved Is a Penny Earned.") Rose went on to give the city an administration marked by "personal liberty" and political corruption. TMER&L enjoyed its respite from persecution. A 1901 stock prospectus had these cheery words for potential investors: "The operating methods and progressive policy of the Company have gained the confidence and goodwill of the public authorities and of the people of Milwaukee." That claim was a wild exaggeration but, after years of open conflict, there was at least a truce on the battlefield.

Beggs and the Politics of Power

The four-cent-fare case was Henry Payne's last major engagement as TM's political field general. He had always made time for other involvements, including service as William McKinley's Western campaign manager in 1896. McKinley won, giving Payne a good friend in the White House, but the Wisconsinite faced trouble at home. A group of reform-minded insurgents led by Robert La Follette had attempted, with spectacular success, to wrest control of the Republican Party from Payne and his fellow bosses. State Republicans denied Payne a seat at the 1896 national convention held to nominate McKinley; he attended only by virtue of his post as a national committeeman. When La Follette won the governor's chair in 1900, Progressive Republicans were in complete control of the state party's machinery, and Payne was, in a sense, a man without a home.

John Beggs, shown with two lieutenants at a company picnic, did nothing to ease relations between Milwaukee officials and TMER&L.

He soon found more accommodating quarters. William McKinley was assassinated on September 6, 1901, in the first year of his second term, and his successor, Teddy Roosevelt, favored Payne with the ultimate patronage job: a cabinet post as postmaster general. The *Washington Post* expressed surprise that Roosevelt would "bring a trained politician of Mr. Payne's stripe into his official family," but the job had been, wrote Payne's biographer, "the ambition of his life." Although an ethics investigation marred his tenure, Payne remained postmaster general until his death on October 4, 1904, at the age of 60. Flags flew at half-mast over every post office in the country, and a company of mailmen served as pallbearers at his Milwaukee funeral.

In 1902, when Mr. Payne went to Washington, John I. Beggs took sole control of TMER&L, succeeding William Cromwell as president and Payne as resident manager. Beggs had been the power behind the throne for at least five years, but his promotion to the presidency, announced only days after Payne's departure, heightened his profile dramatically. The master builder stepped into his predecessor's shoes as the company's public spokesman and lightning rod. He also inherited Payne's mantle as the man Milwaukeeans loved to hate, and he was clearly up to the task. Beggs was, if anything, even more combative than Payne. "When I get to the point," he once said, "where I cease to arouse a certain amount of active antagonism, I hope my friends will get me out of business."

While Payne thrived on the give and take of politics, Beggs had practically no patience for what he considered the penny-ante posturing of public officials. "To hell with injunctions!," he thundered during the 1900 franchise debate. When aldermen tried to use his request for new crosstown routes as a bargaining chip, Beggs snapped, "Then I don't want any." He sought, or bought, the cooperation of local officials (making regular use of a little black bag for "campaign contributions"), but his sole focus was the creation of a model streetcar utility. Self-confident to the point of arrogance, Beggs was determined to give the people what they needed, whether they liked it or not. His Napoleonic style made him, in his own words, "the most unpopular man in Milwaukee," a role he seems to have relished.

Utility-bashing had not ended with passage of the 1900 franchise. Although they didn't have Henry Payne to kick around any more, the city's reformers continued their campaign against TMER&L. That campaign gathered momentum with the arrival of a new force in Milwaukee's public life: the Social Democratic Party. Since its inception in 1898, the socialist organization had shown impressive gains at the polls, contending seriously for the mayor's office in 1904 and adding nine new faces to the Common Council. One of the broadest planks in the Social Democratic platform called for municipal ownership of utilities. Arguing from both economic and ideological grounds, the party contended that it was absurd for a public transit system to serve the interests of a private monopoly. But Milwaukee's socialists were gradualists, not revolutionaries.

With the active support of Milwaukee's aldermen, the Milwaukee Northern Railway provided genuine competition for TMER&L in the northern reaches of the city. (Milwaukee County Transit System)

The "cooperative commonwealth" of their dreams would come eventually, but it would come only after a long period of education and organization. In the meantime, the Social Democrats were pledged to decent representative government, and their obsession with public works earned them the sobriquet "sewer socialists."

Municipal ownership was out of the question, at least until TM's franchise expired at the end of 1934, but there were other avenues to explore. The Common Council's reform coalition — a loose alliance of Social Democrats, Progressive Republicans, and liberal Democrats — enacted a series of new safety regulations, service standards, and other niggling ordinances designed to harass TMER&L. The reformers made their most effective thrusts in 1906 and 1907, when the Council awarded franchises to a pair of new transit firms: the Milwaukee Northern Railway, and the Chicago & Milwaukee Electric Railroad. Both were interurban lines that promised head-to-head competition with TM on regional as well as city routes.

Organized in 1905 by a Detroit engineering firm, the Milwaukee Northern was the more substantial of the two upstarts. The firm's entrepreneurs planned to run cars from a terminal at Fifth and Wells north to Cedarburg, where their line would split into two branches, one serving Sheboygan and the other Fond du Lac. The Fond du Lac branch was stillborn, but the new company was offering hourly service from Milwaukee to Port Washington (the site of the Northern's power plant) by the end of 1907, and the line was extended to Sheboygan in September, 1908. The Milwaukee portion of the line was easily the busiest. Running through neighborhoods already served by TM's trolleys, the Northern offered newer cars and a three-cent fare besides — one of the conditions of its franchise. The new line also offered an unusual service to its rural customers: Passengers who wanted to stop a train at night were instructed to "Stand Clear of Track and show a light or strike a match."

The second upstart, the Chicago & Milwaukee Electric Railroad, was planned as the north-south extension of a small Waukegan line. A.C. Frost, the company's guiding light, pushed in both directions at once, reaching south to Evanston in 1899 and north to Kenosha in 1905. Continuing their northward progress, Frost's crews reached the outskirts of Milwaukee at the end of 1907, when a recession caught up with them. The C&ME Railroad went bankrupt in January, 1908, but receivers soon completed the line to its Milwaukee terminal at Second and Wisconsin — only a block from TM's Public Service Building.

John Beggs, understandably, considered the presence of bona fide competitors an outrage. The Frost line paralleled his own Milwaukee-to-Kenosha route, and the Milwaukee Northern pre-empted a major portion of his long-sought link with the short Appleton interurban he controlled. The new lines posed an obvious threat to Beggs' master plan and, to add insult to injury, both siphoned off lucrative short-haul traffic in the portions of the city they served. In the prevailing political climate, however, Beggs could do little more than grumble.

The climate was about to change. Since the late 1890s, an unlikely combination of civic reformers and utility leaders across the country had been working to take utility regulation to a higher level. The existing pattern of municipal control frightened both camps — reformers because it was often corrupt and utilities because it was often capricious. In a 1907 speech to a Milwaukee men's club, John Beggs stated the case with characteristic bluntness:

I do not think that any of the large public service corporations are afraid of honest regulation. What we are most afraid of is that the present frenzied public opinion will bring about a condition of affairs in which the corporations will receive but little consideration. We are afraid of the regulation of the demagogues of a common council who have developed a hatred toward all corporations and have gone forth as the "saviors of the dear people."

In one of those convergences of opinion that occurs perhaps once in a lifetime, most of Wisconsin's civic leaders agreed that the answer was state regulation, which satisfied both the progressive interests of the reformers and the practical interests of the utilities. The state legislature had established a non-partisan, three-man Railroad Commission in 1905 to oversee the steam railroads. In 1907 the agency's mandate was broadened to include every major utility: water, gas, heating, telephone and, of course, electric. The Railroad Commission (now the Public Service Commission) was empowered to set rates of return, award new franchises, establish service standards, adjudicate disputes, and approve borrowing plans — everything, in short, that municipalities had done and a good deal more. Wisconsin thus became, with Massachusetts and New York, one of the first three states to adopt a system that quickly became the national standard.

Milwaukee was on the losing side of the regulation debate. Local officials were vexed to find TMER&L one large step removed from their control, but that didn't keep them from trying. The City of Milwaukee deluged the Railroad Commission with complaints about the system's safety, service, and fares. One socialist alderman seriously proposed revoking TM's franchise because the company had not equipped its cars with air brakes by the agreed-upon deadline. Another request called for a reduction in streetcar fares from four cents to three. Although it rarely ordered major changes, the Railroad Commission, amply funded by the legislature, investigated each case in detail, sometimes with interesting results. One traffic study found that Milwaukeeans, for some reason, took nearly twice as long to board streetcars as their counterparts in Indianapolis, St. Paul, and St. Louis. A 1907 audit brought an unsolicited compliment from an accountant hired by the Commission:

From the standpoint of the Stockholder, the books disclose results which mark the management of this Company as being of uncommon capacity. From the same standpoint, there is left little to be desired, since in addition to the payment of good dividends, the property has been so well kept up that it is probably in better condition at the end of each year than ever before.

John Beggs considered the report "the highest compliment that had ever been paid to [my] administration of the property."

The anti-TM agitation that marked the Payne-Beggs era reached its final crescendo in 1910, when Milwaukee voters elected Emil Seidel, a Social Democrat and a patternmaker, their mayor. In contrast to the slippery opportunism of David Rose, the socialists had earned a reputation for moral probity, sound planning, and a sober approach to government; the mayor's office and a Common Council majority were their rewards. The immediate reaction in some quarters was alarm. The editors of the *Electric Railway Journal* (Apr. 23, 1910) tried to allay their readers' fears: "Undoubtedly the Socialists would be glad — if it were at all possible — to attempt to inaugurate their hobby of municipal ownership in Milwaukee, but this is out of the question under the present law...."

Emil Seidel, the patternmaker who became Milwaukee's first socialist mayor in 1910

(Milwaukee Public Library)

The transit system may have been beyond their reach, but Seidel and his comrades soon began to build an electric lighting plant powered by the municipal garbage incinerator. (Its 600 kilowatts were no threat to TMER&L.) Daniel Hoan, the new city attorney (and later mayor), filed multiple suits against the utility, forcing TM to pay old license fees and to sprinkle water between its streetcar tracks. Beggs, of course, took vigorous exception to the course pursued by Milwaukee's new leaders. "They have shown," he trumpeted, "an utter lack of the experience and intelligence necessary to administer this government. They have been given rope enough and they have hanged themselves." Good government, in Beggs' view, was temporarily restored in 1912, when the Republicans and Democrats ganged up as "nonpartisans" to deny Seidel a second term.

The Once and Future President

Local politics played a major role in John Beggs' first presidency, and corporate politics played the decisive role in its termination. The unraveling of his power began in about 1903, when the North American Company started to purchase lighting and traction companies in St. Louis. The holding company turned naturally to Beggs in its attempt to create a unified utility. By 1905 he was splitting his seven-day weeks between Milwaukee and St. Louis, logging thousands of miles on the train, and by 1907 the work of consolidation was substantially complete. The St. Louis system was twice as large as Milwaukee's but only half as profitable, providing just the sort of management challenge Beggs relished. He developed close ties in the city, joining several local boards and giving his daughter's hand in marriage to Richard McCulloch, son of a St. Louis streetcar executive.

James Campbell, the St. Louis tycoon who took control of the North American Company in 1909 and gradually pushed John Beggs aside

Beggs also made the acquaintance of James Campbell. Campbell was a central figure in "the Big Cinch," a group of prominent businessmen who dominated civic affairs in St. Louis. He was also a full-time capitalist. Beginning with a small stake in the Southwestern Pacific Railroad, Campbell had branched out into real estate, banks, gaslight companies, and streetcar lines. Dubbed "the J.P. Morgan of the West," he was a multi-millionaire by 1900. Campbell controlled a number of St. Louis traction firms, and he worked directly with Beggs in consolidating them under North American. Within a few years, he became a formidable presence on the holding company's board.

Campbell was, like Beggs, Scotch-Irish by birth, poor by background, and acquisitive by nature. Despite their similarities (or perhaps because of them), the two men developed a cordial dislike for each other. They also differed fundamentally on utility policy: Beggs was clearly a transit specialist, while Campbell favored a more diversified load. The pair found themselves locked in a power struggle that was played out in the North American boardroom. As a director, shareholder, and key executive, Beggs was a dominant force on the board, but Campbell acquired a bigger stake, and a bigger stick.

He soon found an opportunity to use it. The pivotal issue in the struggle between Beggs and Campbell was an extension of Milwaukee Light, Heat & Traction's interurban network. Beggs had been seeking a rail connection with Chicago for years, and his longtime target was A.C. Frost's company, the Chicago & Milwaukee Electric Railroad. As early as 1899, the board had directed Beggs to pursue "the Frost line ... with diligence." When the line went bankrupt in 1908, Beggs, through MLH&T, quietly began to buy up its securities. His word had always been North American's bond, but no longer. In May, 1909, the board, at Campbell's insistence, censured Beggs for unauthorized use of company funds and ordered him to repay the full amount out of his own pocket.

Beggs Isle, a lavish six-acre estate on Lac la Belle, gave John Beggs a presence in TMER&L territory even after he left the company.

The Frost railway would later form the backbone of Samuel Insull's fabled North Shore line, but Campbell wanted no part of a traction empire.

The St. Louis capitalist did not stop there. The flexible federal style practiced by North American's part-time officers practically invited a takeover, and Campbell took over. In December, 1909, when Charles Wetmore developed health problems, James Campbell succeeded him as the holding company's president. John Beggs was relieved of his duties in St. Louis, with a corresponding cut in salary. Campbell began to replace Milwaukeeans on the board with St. Louis residents, and in 1910 he sent his protegé, James Mortimer, to Milwaukee as vice-president and secretary of both TMER&L and MLH&T.

Beggs responded coolly. He did not abandon his manifold interests in St. Louis. In 1910, in fact, he purchased the St. Louis Car Company, which had manufactured hundreds of trolleys and interurbans for the TM system, and installed his personal assistant, Edwin Meissner, as manager. (Meissner bought the company, at Beggs' full asking price, in 1925.) Nor did the aging capitalist abandon his Milwaukee connections. In 1911 Beggs made the one extravagant purchase of his life: an elegant home on a six-acre island in Lac la Belle, only a mile from the interurban station in Oconomowoc. Beggs Isle remained his summer home until his death, and his family's vacation rendezvous until well after World War II.

John Beggs had staying power to spare, but James Campbell gradually wore him down. In February, 1911, North American cut his TM salary from $30,000 to $20,000. On April 1,

14 years to the day after he took over the system, Beggs resigned. Campbell himself stepped in immediately as TMER&L's president, and James Mortimer moved up to the general manager's post. Noting that he was retiring "to give his entire time" to the St. Louis Car Company, the board thanked Beggs for his "untiring devotion, conspicuous capacity and brilliant leadership."

Beggs left the field gracefully. TM employees filled the Public Service Building's auditorium to overflowing for a farewell reception that Campbell and Mortimer managed to miss. After a serenade by the employee band and the presentation of gifts (a painting of a horse, a gold-headed cane), Beggs addressed the assembled throng:

> *Not to be moved almost beyond power of expression by this testimonial would require a nature deeper than that with which I have been endowed. I prize more highly this testimonial which shows that my memory is inscribed in the hearts of my fellow workmen of this great company than I would prize a decoration from any king or potentate on earth.*

His memories of the political side of his tenure were significantly less pleasant. In a retirement-day interview with the *Milwaukee Sentinel*, Beggs, characteristically, had the last word:

> *It is a satisfaction to me to feel that ... after all these years of unreasonable criticism, abuse and misrepresentation to suit politicians and for the purposes of political capital, with a clearly defined principle and system outlined in mind 14 years ago and consistently pursued, there never was a time when the property or myself stood as well as both do today.*

Beggs was clearly entitled to a profound feeling of accomplishment. The Milwaukee Electric Railway and Light Company was, by 1911, his creation and his alone. He had fashioned its policies, its procedures, and its program as a tangibly personal extension of his own character. "A new type of man," wrote the *Sentinel*, "must fill his place." That man was James Mortimer, who would quickly set a course of his own. But Mortimer, and Milwaukee, had by no means seen the last of John I. Beggs.

With Beggs out of the picture for the time being, the company he had created began to march to a different drummer.

Chapter 4

1911-1920

Looking for Load

If the 1896-1911 period had been, in John Beggs' view, "one continual, triumphant march," the next decade resists such easy definition. TMER&L continued to move forward at a brisk pace, but the march of progress split into two distinct columns: electric transit, and electric light and power. Under Beggs, of course, TM had been primarily a transit company. Under James Mortimer, its focus shifted almost instantly to the lighting and power markets that Beggs had virtually ignored. Mortimer did not neglect the traction side — the railway system, in fact, created more than its share of headaches — but the keynote of his presidential term was a pervasive, permanent shift in the balance of power.

The Electric Company

John Beggs and James Mortimer were as different as oil and water. Where Beggs had been a self-taught jack-of-all-trades, Mortimer was a trained electrical engineer, the first of many who would hold the president's office. Where Beggs had been a micromanaging autocrat, Mortimer was a delegator. All 12 department heads had reported directly to Beggs; his successor cut the number to three. Where Beggs had been an empire-builder, Mortimer was a developer, painstakingly filling in the gaps the traction king had left behind. Where Beggs had been a rigid, thoroughly Victorian boss, Mortimer proved himself to be a refreshingly progressive employer. The multiple differences between the two men were, in part, generational. Beggs was 63 when he left the presidency, a ripe old age in 1911. Although a portly frame and a neat black mustache added years to his appearance, Mortimer was only 31 when he took office. A patriarch had been replaced by a whiz kid.

Like most utility executives of his era, James Mortimer had served a nomadic apprenticeship. Born in Elmhurst, Illinois, in 1879, he earned his engineering degree from the University of California, graduating as valedictorian in 1900. After two years of teaching at his alma mater, Mortimer moved on to operating posts at utilities in Seattle and Tacoma, where he caught the attention of Electric Bond and Share, General Electric's holding company. GE hired him to manage other utilities in the West, including the Telluride Power Company in the mountains of Colorado. James Campbell had a financial interest in the Telluride utility, and he developed a personal interest in James Mortimer. Campbell hired Mortimer away from GE in 1909. He made the young man a North American vice-president almost immediately, and placed him in charge of the Milwaukee and St. Louis properties as soon as

James Mortimer, the 31-year-old whiz kid who shifted TMER&L's emphasis from traction to power

John Beggs had been eased out.

Among his other abilities, James Mortimer was a shrewd judge of talent. His most important recruit was a man five years his senior, Sylvester Bedell Way. As a manager who came up the hard way, S.B. Way was, in some respects, a throwback to John I. Beggs. Born in Philadelphia (Beggs' hometown) and raised in Kansas, he spent his early career as a field hand, laborer, and job printer. A scholarship finally took him to Philadelphia's Drexel Institute. With a hard-won engineering degree in hand, Way went to work for a storage battery company whose customers included a St. Louis utility. An installation project there led to a job offer, and by 1903 Way was chief electrician for the entire North American system in St. Louis. In November, 1911, Mortimer brought him up to Milwaukee. (Although the St. Louis system was considerably larger than TMER&L, it served as a sort of farm team for the Milwaukee operation.) S.B. Way was placed in charge of the long-dormant power and lighting side of the company, and he quickly earned a reputation as a tireless worker and a demanding boss. A new slogan was heard in the electrical departments: "Where there's a Way, there's a will."

Sylvester B. Way, Mortimer's right-hand man and a dominant influence on the company for more than 30 years

Both new managers — Mortimer and Way — soon had substantially larger responsibilities. On June 12, 1914, James Campbell died at his summer home in Greenwich, Connecticut. ("Mr. Campbell was noted for his brevity of speech," wrote James Mortimer in tribute. "His methods of approach were decidedly direct.") Campbell's protegé immediately became his successor. In 1914 Mortimer moved to New York City as president of the North American Company, and Way moved up to a new post as general manager of the Milwaukee system. Mortimer remained the man in charge, launching programs and setting policies but, on a day-to-day basis, the Milwaukee utility was Way's to run. His salary was $15,000 a year; Mortimer, who also drew a North American paycheck, received $9,000 annually from TMER&L. S.B. Way would remain the system's heartbeat for the next 32 years.

Mortimer and Way shared one overriding objective: to correct what they perceived as a basic imbalance in the business. John Beggs had operated on the nineteenth-century premise that utilities required a dual load: traction during the day and lighting at night. He had made electricity available to those who wanted it, but Beggs' heart clearly belonged to the trolleys. Mortimer and Way, by contrast, went looking for load wherever they could find it, and the times were propitious. Turbine technology had reduced the cost of generation, bringing electricity within the financial reach of the masses. Manufacturers were turning out a dazzling variety of machines and appliances to put power to use, and the potential market was growing rapidly. Between 1910 and 1920, a time of general prosperity punctuated by World War I, Milwaukee's population jumped from 373,857 to 457,147, a 22-percent increase, and other communities in the region were growing even faster.

TMER&L's new leaders were determined to reach everyone, and one of their first steps was the easiest. Mortimer apparently decided that, as a corporate moniker, "The Milwaukee Electric Railway and Light Company" was both cumbersome and increasingly inaccurate. By mid-1911 the utility was advertising itself as simply "The Electric Company," with its formal name in smaller print. "TMER&L"

would remain the name of record until 1938, but "The Electric Company" signaled a broader intent.

It was S.B. Way's job to put the load-building program into practice, and his staff worked simultaneously in three markets: residential, commercial, and industrial. The proportion of American homes that used electricity more than tripled between 1910 and 1920, rising from 10 to 35 percent, and Wisconsin kept pace with the nation. Lighting, the most frequently requested service, was only a foot in the door. TM advertised the virtues of "The All-Electric Home," and the retail store in the Public Service Building expanded its offerings dramatically. Irons and fans had been mainstays for years, but the company also sold thousands of toasters, percolators, hotplates, night lights, sewing machines, curling irons, lamps (with both fringed and "art glass" shades), cigar lighters, vacuum cleaners, washing machines, hair dryers and, by 1916, kitchen ranges. All, needless to say, consumed more power than the average light bulb. The appliances sold during one 1915 flat-iron sale represented a load equal to that of the entire Racine street railway. TM's involvement did not end with the sale. The company repaired all appliances, installed the larger ones and, as a public service, replaced burned-out light bulbs free of charge.

The residential load practically doubled between 1912 and 1917, but Way and company kept up the pressure. In 1917 an "Electric Santa Claus" toured the interurban towns by rail, pitching everything from toasters to teapots. Perhaps the most imaginative sales tool was *Cinderella Today*, a silent film produced by the company in 1917. With the help of appliances furnished by her fairy godmother, Cinderella said farewell to drudgery and won her prince's heart, not with a lost slipper, but with "a pan of delicious biscuits" baked in an electric range.

The commercial market received just as much attention. Local businesses had been using electricity since about 1880, when the first arc lights flickered on in downtown Milwaukee. Some of the larger hotels and retail buildings had become small-scale utilities, installing generators to supply themselves and their neighbors with light and power. Between 1913 and 1916, TMER&L quietly bought out its remaining competitors in downtown Milwaukee: the Wells, Plankinton, and Commonwealth power companies. At the same time, TM salesmen tried to liberate other merchants from what one ad called "The Waste and Worries of an Independent Plant." Offering cheaper, more reliable power than they could provide for themselves, the company extended service to hundreds of department stores, office buildings, movie theaters, hotels, and other establishments.

Where businesses clustered — in the region's downtowns and neighborhood shopping strips — TMER&L offered "White Way" lighting, featuring 300-watt Mazda lamps with durable tungsten filaments. Until the City of Milwaukee developed its own street-lighting system in 1916, illumination was generally the province of local merchants' groups. In 1913, for

A new focus and a new name: Vehicles like this early electric service truck became rolling signboards for "The Electric Company."

Whether they were calling attention to a nickelodeon on Third Street (left) or ladies' hats at Espenhain's Dry Goods Store on Fourth and Grand, electric lights became a nearly universal feature of commerce.

instance, TM sales engineers designed a White Way system for the Green Bay Avenue Advancement Association. The brilliant lights called attention to entire shopping districts, but some merchants wanted to outshine their neighbors. The number of electric signs in the Milwaukee area rose from a dozen in 1910 to 200 in 1914 and more than 600 in the early 1920s. Gimbels Department Store outdid them all in 1916, installing a 20,000-candlepower searchlight on the roof of its downtown store.

Mortimer and Way did not overlook the commercial market for steam heat. Once a waste product of the system's power plants, steam became a small but important business line, particularly in downtown Milwaukee. In 1917 the company installed a steam tunnel connecting the Commerce and East Wells (Oneida Street) Power Plants. The tunnel extended the steam system's reach and ensured its continuity of supply, enabling TM to scrap the generators in the Public Service Building's basement. The number of steam customers climbed from 230 in 1909 to 819 in 1919, and they included practically all the leading downtown businesses. TM's service was so dependable that new landmarks, including Northwestern Mutual's 1914 home office, were built without independent power plants or boilers.

As important as the residential and commercial markets were, it was the industrial load that TMER&L coveted. Milwaukee was a mature industrial center by 1910. More than half its labor force was engaged in manufacturing, one of the highest proportions in the nation, and civic boosters had dubbed Milwaukee "The Machine Shop of the World." Local industries had entered the electric age in 1882, when Henry Jacobs sold brewer Frederick Pabst his pioneer Edison lighting system, but the lighting load was minuscule in comparison with the market for industrial power. By 1900 steam engines had begun to seem old-fashioned in comparison with electricity, and up-to-date manufacturers were quick to make the change. In practically every case, however, they installed their own power plants; central-station service was perceived as too expensive and too unreliable. In Milwaukee and elsewhere, industry offered a huge potential market. Fewer than 20 percent of America's factories were electrified by 1910, but they generated nearly half the nation's power.

The clear challenge was to replace private power with public or, as one TMER&L publication put it, to "administer knockout drops to gas engines and steam plants" that drove independent plants. TM's generating costs were on the decline, and James Mortimer adopted a rate system that gave large users additional breaks. But local industrialists still needed convincing. Roving teams of engineers visited plants throughout the region, analyzing power consumption in minute detail and comparing the costs of central-station service with those of the manufacturers' own plants. Progress was often slow, but the rewards were almost instantaneous: A good-sized factory could consume as much power as a small city. In 1913 Crucible Steel became one of the first foundries in the country to install an electric arc furnace. Able to liquefy 1.5 tons of metal in a single heat, the furnace used enormous quantities of energy, but the device paid for itself within a year. By 1916 there were six electric furnaces in local foundries, and they constituted the largest single class of users in the entire system.

TMER&L did not overlook its smaller industrial prospects. Even the tiniest machine shops were courted and, once they'd signed on, the company sold them motors, transformers, soldering irons, embossing dies, and other electrical appliances. A Milwaukeean could buy a toaster for his home and a glue pot for his shop from the same retail store in the Public Service Building.

World War I carried the demand for industrial power to new heights. The American economy slowed down when the shooting started in 1914 but, by the time the United States entered the conflict three years later, the nation's factories were working overtime. Milwaukee did its part, supplying the Allies with everything from canned meat to marine gears; the city's industrial output jumped 158 percent between 1915 and 1920. As new factories were erected and old ones were expanded, private power plants were overwhelmed, opening the door for TM's sales engineers. By 1919, when "machine use" accounted for 75 percent of the system's peak load, manufacturing had outstripped residential users, commercial customers, and even the street railways as the region's largest consumer of electrical power.

Under the Mortimer-Way team's leadership, TMER&L concentrated its load-building efforts on the Milwaukee area, penetrating the market block by block, house by house, shop by shop. Although he was not an expansionist in the Beggs mold, Mortimer was also alert to opportunities outside Milwaukee. In 1912, for instance, North American purchased the Kenosha Street Railway from its English owners. (The holding company already controlled Kenosha's electric utility.) The purchase helped to trigger the formation of a new North American subsidiary: the Wisconsin Gas & Electric Company, incorporated on September 20, 1912, with S.B. Way as president. Sometimes referred to as TM's "little brother," WG&E furnished light, power, and transportation in Kenosha and gas service in Racine. Racine's power plant and street railway remained, for unspecified reasons, under TMER&L's wings.

The gashouse on Racine's lakefront was Wisconsin Gas & Electric's most important asset by a wide margin. Electric utilities and gas companies had been mortal enemies in the late 1800s, when both were competing for lighting customers, but North American, after some hesitation, had seen the wisdom of diversification. A private syndicate led by John I. Beggs had purchased the Racine Gas Light Company in 1899. The partners put up a new plant on the lakefront and finally convinced North American to buy the operation in 1902. The Racine works supplied "manufactured gas." Soft coal was heated to a red glow in the absence of

Heavy industries like the Bucyrus-Erie Company in South Milwaukee represented the greatest potential market for electric power. Fifty-two men could fit in the dipper of this B-E mining shovel. (Milwaukee Public Library)

The gas plant on Racine's lakefront produced gas for heating, cooking, and heat-treating. Its most abundant by-product — coke — became a popular foundry fuel.

oxygen, producing solid coke for industrial use and flammable gas for heating, cooking, and (for a few more years) lighting. By 1914 WG&E's gas lines had been extended south to Kenosha and north to South Milwaukee, where nearly 600 people attended the opening celebration. By 1919, when the plant served 21,825 customers, Racine-made gas was generating more than $1 million in gross revenue every year.

Although gas was the new company's reason for being, the "Electric" in Wisconsin Gas & Electric was not overlooked. Fully 75 percent of Racine's industries had converted to "The Central Station Idea" by 1915, and Kenosha's customers included the brand-new Nash automobile plant. WG&E also displaced Milwaukee Light, Heat & Traction, the old interurban subsidiary, as the TM system's rural arm. The company's sales agents (11 by 1916) scoured the countryside around Milwaukee, buying up independent utilities and tying them into the regional system. In 1912, for instance, WG&E bought the Burlington Light and Power Company, whose hydro plant on the White River added a grand total of 125 kilowatts to the larger network. From Eden in Fond du Lac County to Johnson Creek in Jefferson County to Paddock Lake near the Illinois line, the company opened a broad crescent of territory surrounding Milwaukee. In 1916 alone, line crews stretched 169 miles of wire in southeastern Wisconsin, reaching Whitewater, Lake Mills, and Theresa. Although it was modest in comparison with the all-out campaigns of the 1920s, WG&E's load-building program placed it at the forefront of America's rural electrification efforts.

On every front — city and country, homes and shops, foundries and factories — the TM system's attempt to shift the balance of power was a resounding success. Between 1909, on the eve of James Mortimer's arrival, and 1919, his last full year as president, the number of electric customers connected to the system's lines soared from 11,567 to 85,030 — an increase of 635 percent. Power consumption grew even more rapidly, jumping from 21,859,000 kilowatt-hours in 1909 to 237,696,000 a decade later. The system's load nearly doubled every three years! But the power shift is best measured in terms of how TM's power was used. In 1906, near the high point of John Beggs' reign as the region's transit king, electric railways consumed 74 percent of the company's generated output. In 1912 they used 54 percent, and by 1917 the figure had fallen to 26 percent. In less than a decade, The Milwaukee Electric Railway and Light Company became, in fact if not in name, the Electric Company.

Trouble on the Tracks

Despite exponential growth on the electric side, TMER&L remained *the* transit company in southeastern Wisconsin. Annual patronage never fell below 129 million passengers during James Mortimer's tenure, an average of more than 350,000 rides per day. (A less obvious statistic from 1916 underscored the system's importance: Passengers left a total of 2,800 umbrellas on TM's cars during the year; only 1,600 were reclaimed.) Although their priorities clearly lay elsewhere, Mortimer and Way never treated the transit system as a stepchild. New routes were opened, new car stations were built, and a steady procession of new cars emerged from the Cold Spring shops. The number of passenger cars in service rose from 477 in 1912 to 716 in 1919.

But disturbing trends were apparent. Between 1896 and 1910, TMER&L had averaged a 27-percent increase in ridership annually. Patronage continued to rise between 1910 and 1920, but the average annual growth rate fell to 6 percent. Milwaukee's experience reflected a national slowdown. Between 1907 and 1919, new trackage on America's street railways dropped from 1,880 miles to 141, and the number of new cars on order declined from 6,216 to 2,447.

Several forces acted as brakes on the transit system's growth, including what might be called the curse of familiarity. In 1890 electrified transit had been hailed as a world-changing breakthrough. The world did in fact change, but its residents, as always, adjusted easily. Two short decades after its Milwaukee debut, the streetcar was viewed as basic public transportation, a ubiquitous civic necessity with all the glamour of a waterworks or a sewage plant.

A fresh alternative had emerged to satisfy the public's appetite for novelty: the automobile. Its grip on the American people began earlier than might be supposed. At the turn of the century, automobiles were little more than expensive toys. (S.B. Way, interestingly, owned one of Milwaukee's first, an "open-air electric," and he remained an avid motorist all his life.) Wisconsin did not require motor vehicle licenses until 1905, when drivers were charged

The shape of things to come: Automobiles and streetcars jostled for space in downtown Milwaukee. The view is east from Water and Wisconsin.

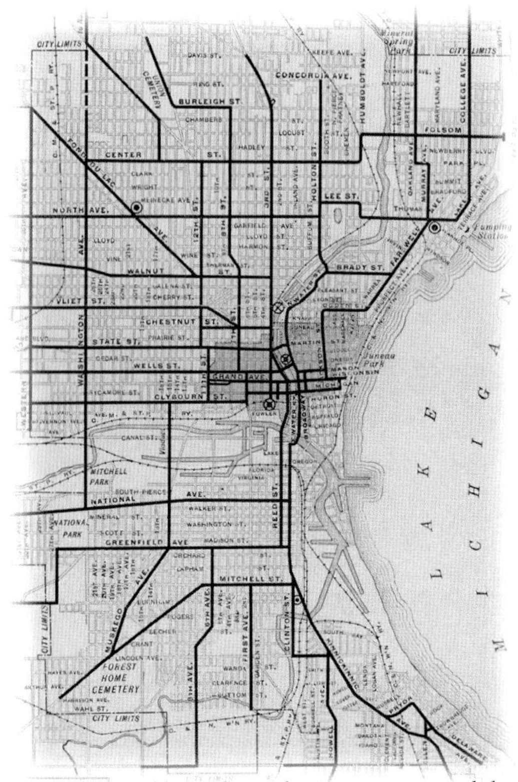

Despite trouble on the tracks, TMER&L provided essential transit service. Milwaukee's streetcar system reached every corner of the city by 1913.

one dollar for a lifetime permit. In 1910 there was only one automobile for every 84 families in the state. Then came the deluge: By 1920 there was one for every two families. The growing legions of motorists quickly demanded new roads for their new vehicles; the state trunk highway system mushroomed from less than a mile in 1912 to nearly 250 miles in 1920, and Milwaukee County alone boasted 160 miles of concrete roads at the decade's end. Car-buying and road-building would reach epidemic proportions in the 1920s, but the automobile, not the streetcar, was already the vehicle of choice for most TMER&L customers.

The interurban network was first to feel the impact of the automobile. However regular its schedules, however reasonable its fares, no train could begin to compete with the convenience of the private car. Milwaukee Light, Heat & Traction, the interurban subsidiary created by John Beggs in 1896, carried nearly 9 percent of the entire transit system's riders in 1912. By 1920 its proportion of the traffic had fallen to just over 2 percent. The subsidiary's financial performance was even more troubling: MLH&T showed a scant 0.71-percent return on its investment in 1918, less than a tenth of its target.

On February 1, 1919, as the system's interurban arm continued to wither, MLH&T and all its trains were quietly absorbed by TMER&L. The utility was once again a two-company system: TMER&L in Milwaukee, Racine, and the interurban towns; and Wisconsin Gas & Electric everywhere else in southeastern Wisconsin. ("System" and "company" are used interchangeably in this narrative; WG&E was, in effect, the regional arm of TMER&L, and both firms were managed by the same executives.)

Bound for Wauwatosa, Car 203 paused in the "country" near 60th and Wells Streets.

A Cudahy suburban car rumbled down Plankinton near St. Paul Avenue. (Russell Schultz)

City lines fared better. The automobile was a growing presence (collisions between streetcars and gasoline vehicles increased 140 percent between 1914 and 1915), but ridership continued to grow at a modest pace. It was not passengers, however, that James Mortimer counted, but profits, and he found the transit system wanting. In 1912, according to a Railroad Commission study, the transit side represented 70 percent of TM's total capital investment. In the same period, transportation employees (motormen, conductors, and shop workers) made up more than half the company's labor force. The transit system generated 63 percent of the company's net income in 1912 — a proportionate share — but the system's profits deteriorated as ridership stagnated and expenses rose. Net income from traction operations actually dropped 18 percent between 1912 and 1917, from $692,000 to $567,000. The power and light business, in the meantime, had taken off. Its net income soared, passing the transit operation's in 1914 and steadily pulling away. By 1917 the electric side generated 57 percent of TM's net income — nearly reversing the ratio of five years before — and it did so with fewer employees than the transit system and perhaps half its capital investment.

It should be stressed that TM was not losing money in the transportation business — not yet, at least. It was simply not making enough money to satisfy its investors. The Railroad Commission had established 7.5 percent as a "reasonable" return on the company's investment. In 1918 TM's city lines showed a return of 4.36 percent, while the electric side generated 5.58 percent. That disparity soon became a chasm: By 1921 transit operations returned 4.6 percent and the electric utility generated 7.76 percent. James Mortimer had noticed the trend during his first years in office. As early as 1914, he was complaining that "the street railway utility in Milwaukee ... is suffering from absence of growth of revenue over which the owners have no control." In a 1915 interview with the *Electric Railway Journal,* he said, "This company's investment in the railway business is now earning so little as to make the expenditure of additional capital for further facilities absolutely impossible without obtaining some relief."

Mortimer, S.B. Way, and their traction lieutenants did what they could to improve the bottom line. Routes were extended voluntarily only when potential customers, notably outlying

Track reconstruction closed Grand (Wisconsin) Avenue to traffic in 1913. (Milwaukee Public Library)

factories, offered what the board called "the most advantageous terms obtainable," i.e., subsidies. Old streetcars were rebuilt as trailers to save money. New cars were built with economy in mind; the 800-series model, which debuted in April, 1920, weighed only 15 tons, 10 less than some of TM's older cars. Freight service was added to all the interurban lines between 1915 and 1922. As a sign of things to come, freight trucks began to run from the railheads at Watertown and Burlington in 1919, and the company authorized an experiment with "automobile bus" service to Menomonee Falls in the same year. On April 17, 1920, the system's first city buses began to operate between 11th and Mitchell and 31st and Burnham; the service was described as "an extension which would not justify the use of more streetcars."

Such measures brought in revenue, but the company's options were severely limited. If TMER&L had been a genuinely private business, its owners could have raised fares and/or cut expenses significantly to meet the perceived revenue shortfall. As a regulated utility, however, TM could do neither, which rankled James Mortimer no end. Rate relief was nearly impossible. The Railroad Commission turned down repeated requests for fare hikes, because, Mortimer charged, they were politically inexpedient. In 1912, in fact, the Commission reduced fares marginally, requiring the company to sell 26 tickets for one dollar, rather than 25. Mortimer won a small victory in 1914, when TMER&L became perhaps the first transit system in the nation authorized to charge zone fares;

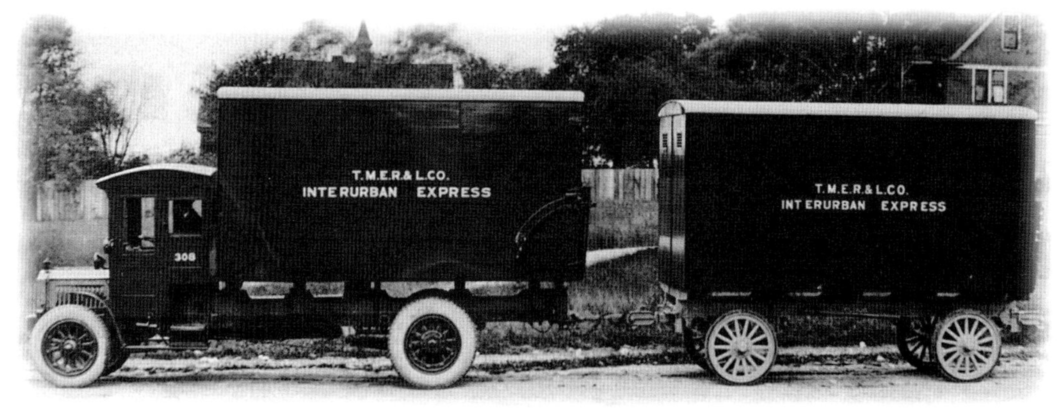

TMER&L began to run freight trucks from its interurban railheads in 1919.

Milwaukee's first buses went into service on Mitchell Street in 1920. The 25-passenger vehicles were actually converted from freight trucks at the Cold Spring shops.

the cost of a ride thereafter bore at least some relation to its distance. In 1918, after a vicious round of wartime inflation, TM was allowed to eliminate the sale of commutation tickets, thereby raising the basic fare to one nickel. The same rate, Mortimer noted wryly, had applied on the horsecars of the nineteenth century. Fares were not the only point of contention; the Madison regulators also forced TM to buy new cars, maintain unprofitable routes, and extend new lines that made little economic sense to the company.

In John Beggs' view, the Railroad Commission had been "the best thing that ever happened to TMER&L," ensuring fair treatment for the company and first-rate service for the public. James Mortimer believed otherwise. "The regulation of public utilities," he wrote in 1917, "as at present practiced is too ponderous and slow to give to the utilities and investors the protection originally intended. Near-bankruptcy has to be shown before relief can be obtained." The Railroad Commission's charter specified regulation, not protection, but Mortimer's frustration is understandable. The overworked commissioners routinely let cases rest for four to five years before they issued a ruling.

The prime instigator in the majority of cases involving TMER&L was the City of Milwaukee. The political turmoil of the Payne-Beggs years came to a temporary halt in 1912, when Republicans and Democrats united behind Dr. Gerhard Bading as their "non-partisan" mayoral candidate. Bading, the former city health commissioner, defeated socialist Emil Seidel, and it appeared that a new day of cooperation had dawned. When a bunting-draped TM

Built with economy in mind, 800-series cars were 10 tons lighter than some of their predecessors. Car 855 operated on the Eighth Street line.

car made its maiden run over the 27th Street viaduct in 1913, Mayor Bading was at the controller, and two years later he served as toastmaster at a company banquet. In the 1916 election, however, the lackluster Bading was defeated by a young city attorney, Daniel Webster Hoan. Hoan would spend his next 24 years in the mayor's office, and TMER&L would spend all of that time as his favorite whipping boy. A colorful campaigner, a shrewd politician, and a devout socialist, Dan Hoan based his utility-bashing on a simple premise: "Such property as is publicly used should be publicly owned." Municipal ownership was his ultimate goal but, in the meantime, Hoan brought the full weight of municipal pressure to bear on the company.

The new mayor's anti-utility campaign had begun during his 1910-1916 tenure as city attorney. After repeated attempts, Hoan had secured the right to enforce an old ordinance requiring TM to pave its track zone (and one foot outside). For a system covering more than 160 miles of track within the city limits, the order was more than an inconvenience. In 1915 Hoan had filed a 225-page case against the utility, alleging "crowded street cars and poor service." The Railroad Commission responded by issuing new service standards and levying a $10,000 fine. Hoan kept up the pressure as mayor. In 1919 the City of Milwaukee decided that TMER&L needed a separate franchise to operate its interurban cars on city streets. When the company refused to apply for one, it appeared that riders on the Watertown, Burlington, and Kenosha lines would have to leave their cars at the city's edge, walk across the

Mayor Daniel Hoan on the stump. He was probably lambasting the Electric Company. (Milwaukee Public Library)

border, and then climb onto "official" TM trolleys. Such silliness was averted only by action of the Wisconsin Supreme Court.

The only thing that Dan Hoan had in common with TMER&L was a cordial dislike for the Wisconsin Railroad Commission. By 1918 James Mortimer was so fed up with regulatory delays and adverse decisions that he was calling for state ownership of the utility. Hoan, for his part, published *The Failure of Regulation* in 1914, a book-length castigation of the Commission as a shield for corporate greed. "No shrewder piece of political humbuggery and downright fraud has ever been placed upon the statute books," he charged. "It is supposed to be legislation for the people. In fact, it is legislation for the moneyed oligarchy."

It will be noticed that the focus of Hoan's crusade was the transit system, not the electric utility. Power and light were hard targets; Social Democrats might have complained about rates, but the current was, essentially, either on or off. Streetcars, on the other hand, offered highly personal, highly variable service; anyone who had ever been forced to wait in the rain for a crowded car with a grumpy conductor was apt to find at least some merit in Hoan's complaints. The mayor never forgot that trolley lines were lifelines for the working-class voters who kept him in office.

Political pressure on the transit system, combined with general economic stagnation and growing automobile competition, made life difficult for TM's transit managers. In an article written for the *Electric Railway Journal* (Oct. 11, 1919), James Mortimer described in detail the street railway's fall from grace:

> *The special privilege to occupy the streets turned out to be a privilege to pave and clean the streets for the use and benefit of any other kind of transportation. The reputed natural characteristics [of monopoly], which were presumed to charm off competition, never caused Henry Ford a moment of anxiety. The economic regulation that compelled people to ride somehow was supplanted by the legal regulation which compelled the cars to travel whether the people rode or not....*
>
> *It must be confessed that there is no particular psychological appeal in the traction business as now constituted to induce liberal patronage. It cannot adopt as its motto "The more you ride, the more you want to." Time was, when the rail was flanked by cobblestones and wooden sidewalks and a trolley ride was our most popular outdoor sport. Is it possible to restore the old-time zest and flavor? The low number of rides per capita — less than one ride per day — indicates a minimum of patronage, an absence of convenience or pleasure riding and a dearth of salesmanship. We call it "the riding habit," a proper designation. Riding is habitual and forced, rather than attracted.*

"Give the Company Your Loyalty"

If the street railway caused James Mortimer some frustration, another part of the business gave him immense satisfaction: employee relations. Between 1911 and 1920, the system's total employment climbed from 3,125 to 4,874. TMER&L and WG&E were broadening their geographic reach, filling in the gaps between the interurban spokes of the Beggs era, and the effort required people. But expansion had its drawbacks. As the company grew, so did the walls between its operating units. The divisions between the electric utility and the transit system were most obvious, but fragmentation was apparent in other areas as well; a power plant worker in Milwaukee had little reason to associate with a motorman in Kenosha or a lineman in Lake Mills. As the system became increasingly attenuated, Mortimer wanted to make sure that its employees — all its employees — shared a community of interest. The result was one of the most extensive and most sophisticated employee welfare programs in the American utility industry.

John I. Beggs had taken a few tentative steps on the road to better industrial relations. His Public Service Building featured lavish facilities for TM employees, and he supported activities ranging from an all-employee band to the annual picnic at Waukesha Beach. But Beggs somehow never got around to implementing a full-scale welfare program. Some of his policies, in fact, demonstrated a definite lack of trust and respect. Long hours and split shifts were the rule in many departments, but Beggs' most offensive practice was his use of company spies. "We have plain-clothes men constantly about listening to the gossip of our men," he baldly admitted to a reporter in 1909, "not so much for the purpose of ferreting out strikes as to hear what gossip is floating about."

Not surprisingly, James Mortimer encountered a rather high level of employee dissatisfaction when he took office in 1911. A union organizing campaign led to a brief transit strike in April, 1912, and the strike was followed by a legislative investigation of TM's working conditions. State officials found that the trainmen worked "excessively long hours" (more than 10 a day for three-fourths of the men), that they were subject to "the tyranny of the petty official," and that they were routinely spied upon during their off-duty hours.

By the time the investigation closed, James Mortimer had already moved to higher ground. On February 28, 1912, honoring a pledge he had made the previous Christmas, Mortimer launched what would be his most enduring legacy: the Employes' Mutual Benefit Association. The EMBA was, in its original incarnation, primarily a medical plan. A team of full-time physicians, headed by Dr. Charles Lemon, provided free medical treatment for members — in their homes, in a network of regional offices and, for surgical cases, in local hospitals. The infirm and the injured also received a sick benefit of a dollar a day, and those who failed to survive treatment passed on a $300 death benefit to their heirs. (Dr. Lemon was by all accounts a first-rate physician, but his knowledge was subject to the limitations of the time. "Ice-water is condemned by all medical authorities," he wrote in 1916, "and the drinking of tea is said to be responsible for the large number of suicides which occur in Japan.")

As the company and its employees gained experience with the plan, the menu of benefits and services expanded steadily. Visiting-nurse service was added in 1914. Life and off-duty accident insurance policies were offered in the same year, with premiums well below market rates. Medical care was extended to the families of members in 1916, and obstetrical service was added in 1918. The number of house and office calls made by EMBA physicians soared from 3,197 in 1912 to 13,446 a decade later. Free medical care had a profound impact on the health, and therefore the productivity, of the system's employees. The number of sick days per worker dropped from 9 in 1912 to 5.3 in 1913 and 4.9 in 1920 — half the national average.

By 1914 TMER&L employed more than 4,000 workers, including this rough-and-ready crew of pole-painters.

EMBA membership was entirely voluntary. A one-dollar initiation fee and monthly dues of fifty cents paid the doctors' salaries; the company matched employee contributions dollar for dollar, and provided office space and clerical help to boot. The proportion of full-time employees who held EMBA cards rose from roughly 40 percent in 1912 to 75 percent in 1915 and "practically all" in 1918. The Association was especially popular among TM's trainmen, 91 percent of whom had joined by 1914. Although management exercised an ever-present influence, a majority of the EMBA's directors were elected by popular vote in the system's power plants, car stations, and other "districts." James Mortimer strongly encouraged full participation. In a 1913 speech to employees, he spoke fervently of the corporate virtues embodied by the EMBA:

A fistful of lightning: The Employes' Mutual Benefit Association was one of the most high-powered employee welfare programs in America's utility industry.

> *Individuals may come and go, but your Company goes on forever. Give the Company the best you have in you and it will reap its own reward. Give the Company your loyalty and your untiring efforts and see how much easier your work becomes. Give to your associates brotherly love, which has been the foundation of society and human endeavor since the world began.*

From its beginnings in medical care, the EMBA's "brotherly love" grew to encompass practically the entire "people side" of the business, from counseling to sports, from amateur dramatics to vocational education. The Association's secretary, a full-time employee, provided personal assistance in the areas of money management, landlord-tenant problems, estate settlement, consumer complaints, and even marriage counseling. The EMBA sports program provided opportunities for athletes of every persuasion. Baseball

A contingent of EMBA officers dedicated the Oakland Avenue car station in 1917.

was among the first organized team sports, with much of the action taking place on a company-owned diamond just south of the Fond du Lac Avenue car station. Indoor baseball, played with an oversized ball and an undersized bat, came to the Public Service Building's auditorium in the winter of 1913; the room was hung with netting to protect the windows and plasterwork. Golf enthusiasts developed an indoor course in the PSB in 1917. Cold Spring employees laid out two tennis courts near their shop in the same year, giving rise to an active tennis club. Milwaukee being Milwaukee, bowling was the most popular sport of all, and the clatter of falling pins could be heard practically every night on the

The EMBA baseball team played on a diamond adjoining the Fond du Lac Avenue car station.

The indoor squad's home field was the auditorium of the Public Service Building.

PSB's second-floor alleys. Once a year, beginning in 1914, teams from all parts of the service area competed in a system-wide handicap tournament.

Non-athletes were not forgotten. The company picnic at Waukesha Beach, planned by an EMBA committee, was by far the largest social event of the year. In 1913, when 8,000 employees and their families attended, the throngs consumed 600 gallons of ice cream, 25,000 cones, 3,000 packages of candy, 20 cases of lemons, and 3 barrels of sugar. The boat rides, games, diving displays, and refreshments were so popular that the picnic became a two-day affair in 1914, with different groups making the trek each day. By 1919 attendance had climbed to 15,000, and the TM system's picnic was said to be the largest in Wisconsin.

The PSB auditorium was the venue for other entertainments during the year, most provided under EMBA auspices. The company's thespians staged plays like *The Little Rebel* and *A Regiment of Two,* and the troupe donned blackface for an occasional minstrel show. The band and chorus were featured attractions at employee gatherings. Regular dances, including "mixers" held over the lunch hour, drew hundreds of waltz and polka enthusiasts. Most dances were relatively tame affairs, but the company's managers felt compelled to exercise a parental role at least once. In 1914, when the tango was riding a wave of popularity, management banned that rather sensuous exercise from the dance floor, stating only that it had earned "the disapproval of many sincere and earnest persons." Some of TM's younger female employees promptly organized a tango party in the auditorium, with no men invited.

Education became another EMBA

From the brass band that greeted employees ...

... to a tug-of-war that attracted TM's heavyweights, the company picnic at Waukesha Beach offered something for every member of the family ...

... even the small fry.

"department." James Mortimer had launched a training program early in his tenure, offering classes for everyone from linemen to salesmen. In 1919 the effort was formalized and expanded under EMBA sponsorship. Dr. Arthur Rowland, former dean of the Drexel Institute (S.B. Way's alma mater), developed a curriculum that covered blueprint reading, stenography, boiler room practice, business economics, practical electricity, and even poultry breeding. Some courses were compulsory and offered on company time; others (including, presumably, poultry breeding) were strictly electives. Dr. Rowland also developed an "Americanization" program for "alien members" of the EMBA. During the "Red scare" that followed World War I, recent immigrants were considered easy prey for Bolsheviks, anarchists, and assorted other radicals (including union organizers). TMER&L was one of many companies that tried to stem the Red tide with classes in the English language, American civics, and naturalization procedures. "Men so taught," wrote an EMBA official, "make better workers, better men and are less easily led astray by the trouble makers."

The EMBA managed so many activities, from athletics to Americanization, that a regular publication was needed to cover them all. In June, 1913, *Rail & Wire* made its debut as a means "to strengthen the bond of fellowship, the spirit of mutuality between man and man and between man and Company." Sponsored by management in its first few years, *Rail & Wire* became the official EMBA newsletter in 1918. The monthly publication featured articles about sports, classes, entertainment, and noteworthy corporate events, but it was basically a human-interest publication — in other words, a gossip sheet. *Rail & Wire* spread personal news from one part of the system to all the others, promoting a sense of employee belonging and corporate fellowship. In 1913, for instance, a faithful reporter contributed these notes from his department:

Launched in 1913, Rail & Wire *quickly became the system's principal means of communication.*

EMBA variety shows kept employees entertained ...

... while educational programs, including this linemen's school, kept them up to date.

The spring rain and summer sunshine has brought out a most beautiful crop of down on John Wrobbel's upper lip.... M. Unser has returned to work after a ten days siege with tonsillitis.... Anyone desiring information about married life will please call on Reuben Clas — experience two months.

The last addition to the EMBA's responsibilities, and certainly the most novel, was announced in 1918, when the Association became, in effect, a company union. Under a new contract, the EMBA agreed to act as the collective bargaining agent for its members in all negotiations involving wages and working conditions. TMER&L decided, furthermore, to adopt a diluted version of the closed shop; management agreed to hire only EMBA members, and the EMBA agreed to provide the system with all its new non-managerial employees. The agreement was a direct outcome of the labor shortage, and the labor strife, that accompanied World War I. The federal government recognized labor's right to organize and provided, through its War Labor Board, a vehicle for the mandatory arbitration of labor disputes. In its 1918 report to shareholders, TM frankly acknowledged that the EMBA agreement had enabled the company "to avoid the reference of disputes to the National War Labor Board and succeeded in minimizing among its employees, that industrial unrest which has been so common during the last half of the year just closed." Management created an internal union to prevent one from being organized by outsiders.

As the collective voice of its members, the EMBA did sometimes make demands that the company refused to grant, but the Association's role in a highly public 1919 controversy was somewhat less independent. With war production overheating the American economy, the cost of living precisely doubled between 1915 and 1920 — the greatest five-year increase in the nation's history. Workers across the country demanded raises to make up some of the ground they were losing. The problem, in the TM system's case, was that wage increases were felt to be dependent on revenue increases, and revenue increases were dependent on the Railroad Commission. When the Commission failed to act on a request for increased transit fares, EMBA officials, with the full knowledge and tacit approval of management, took their case directly to the state regulators, threatening to strike if TMER&L's request was not granted.

On January 1, 1919, with the Commission showing no disposition to act quickly, the EMBA called a transit strike that crippled Milwaukee and the surrounding region. The trainmen returned to work only after local business leaders had persuaded TM to grant a temporary wage increase. Fueled by inflation, the pressure continued to build. It was not, however, until December, 1919, that the Commission approved a 40-percent increase in the basic transit fare, from a nickel to seven cents. Wages rose accordingly, and peace was restored.

TMER&L's critics charged that the company had manipulated its union, using it as a pawn to push the Railroad Commission into a corner. Legislative investigators chastised the Commission for its failure to "function expeditiously"

The EMBA lodge room, now the kitchen area of the employee cafeteria, was the scene of rituals every bit as esoteric as those of the Masons or the Odd Fellows.

in TM's case, but they also criticized the company for its "failure to meet the just wage demands" of the system's workers. Although officials found no evidence of conspiracy between management and the EMBA, they concluded that the Association's role in the controversy amounted to "coercion."

The Employes' Mutual Benefit Association was a "fraternal industrial union," the publisher of *Rail & Wire,* an educational institution, a social organization, an athletic association, and a medical plan, but it was more than the sum of its parts. To its most devoted members, the EMBA was a way of life. A 1920 *Rail & Wire* editorial declared that the Association was nothing less than "a spirit-forming and spirit-developing society ... that is a new force in our civilization." Just as James Mortimer had hoped, the system's employees made the organization their own and gave it a life independent of management dictates. Like the Knights of Pythias, the Modern Woodmen, and other secret societies, the EMBA developed an entirely distinctive body of ritual. Teams of officers presided over initiations, solemnly sharing the Association's handshake and secret password: "I bring relief." (No one, apparently, brought the Rolaids.) New members could aspire to any number of degrees, including "benevolent inspector" and "master in grand council." The EMBA adopted its own logo, a clenched fist holding five lightning bolts that were said to represent loyalty, honesty, brotherly love, patriotism, and energy. The logo was embroidered on blue-and-gold banners that were unfurled for every gathering. Fellow members were addressed as "Brother" (or, presumably, "Sister;" 445 of the Association's 4,000 members were women in 1919). There was even an official EMBA song, *On, Electric,* sung to the tune of *On, Wisconsin:*

On, Electric, on, Electric, we're your faithful crew.
Through the years to more than twenty we've been part of you.
Through your years of growing glory you've been to us true.
Conquests still wait our efforts, we and you.

The EMBA was, in effect, a poor man's Masons. Quite intentionally, it provided a security blanket that stretched from the cradle to the grave. A member's child could come into the world with the help of an EMBA doctor, grow up in an EMBA household attending EMBA social events, come to work through the EMBA employment office, participate in EMBA sports and classes until retirement, and go to his or her grave with an EMBA escort. The Association's funeral ritual was perhaps the most impressive of all. When a popular motorman died in 1914, a solemn procession of 300 EMBA members, wearing full uniforms and marching in close ranks, turned out for

From cradle to grave, the EMBA was a "spirit-developing society" that had an enormous impact on the lives of its members — and the company that employed them.

his last rites. The EMBA band played the prescribed hymns and dirges, and a white-gloved officer read the EMBA funeral service while the EMBA drill team circled the gravesite in formation.

From pomp and circumstance to parties and picnics, the Employes' Mutual Benefit Association offered something for literally everyone. The Association was so thoroughly enmeshed in the fabric of the TMER&L system that it was difficult, after a few years, to tell where the company ended and the EMBA began. There were, however, other employee programs that flourished outside the shelter of the EMBA umbrella. In 1912 the company established a pension program (with benefits of up to $750 a year) for employees who had reached the age of 60 with at least 15 years of service. A Veterans Association was organized four years later to honor employees who had been on the job for 20 years or more. Profit-sharing began in 1913, when Cold Spring shop hands received extra pay for exceeding production standards. By 1915 the system had been extended to all departments, even Accounting, and by 1920 profit-sharing contributed an extra 11.5 percent to the average worker's paycheck. On April 17, 1914, the Employes' Mutual Saving, Building & Loan Association held its inaugural meeting, with 400 thrift-minded employees in attendance. By 1920 the EMSB&LA's assets had grown to more than $5 million, and the Association had enabled fully 900 employees to buy or build their own homes.

The system's multitudinous employee programs, EMBA-sponsored and otherwise, were expensive to maintain, difficult to administer, and somewhat distracting at times. They were also unquestionably good for the company. Healthier, happier employees obviously performed at a higher level, but every program had an even loftier objective: solidarity. TMER&L was a chronically embattled utility, subject to intense pressure from politicians, regulators, labor organizers, and a frequently dissatisfied public. Comprehensive welfare activities helped to give the company an inner stability that would, James Mortimer hoped, prove impervious to any and every external threat. His public statements, accordingly, often reflected a circle-the-wagons mentality. Mortimer issued this warning to employees in 1913:

Honoring the old-timers: The first Veterans Association dinner was held at the Pfister Hotel in 1916.

> *There are people today who are endeavoring to foment dissension in the ranks of our citizenship, setting up false gods and hopes for your adoration. They travel in the guise of your friends, but at heart are raging wolves.... Lean not toward the enemies of your Company.*

Organized labor was the enemy Mortimer feared most, and he borrowed freely from labor's rich rhetoric to build solidarity in his own system. The president took every opportunity to assert the dignity and worth of the honest workingman. Such sentiments reached a novel height in 1916, when TMER&L published *Our Songs of Labor* as a Christmas keepsake. The book was filled with illustrated poems praising the contributions of everyone from powerhouse coal handlers to messenger boys. This tribute is typical:

Who braves the winter's cold and sleet
And equally the summer's heat,
Climbs swaying poles when winds blow strong
Till he has righted what was wrong?
The Troubleman.

The rhetoric was not without substance. Corporate self-interest may have underpinned the welfare effort, particularly in its formative stages, but Mortimer's philosophy went deeper than union avoidance and other defensive strategies. Like many executives before and since, Mortimer came to see that his was a human business, an enterprise whose success hinged on the conduct, confidence, and character of its people. In his last address to TM employees in 1920, the departing president looked back on the road he had traveled since 1911. His remarks have a surprisingly modern ring:

THE MESSENGER

We see him as he runs about the place
With youthful carelessness; a boyish grace
Is in his manner, smiles upon his face.

Shop, storeroom and office see his way
As sturdily his duty day by day
He interrupts with touches of his play.

Small though he is, forgetful, careless, too.
He stands beside the sturdy and the true—
Trusted as brother worker through and through.

He's but a boy, not yet full seriously
Taking his place. But we who older be
Know in him lies tomorrow's history.

THE CARMEN

To half the city all the time
They are the Company.
The part so many people know,
So many people see.
What they may do, to many folks,
Is Comp'ny policy.

They know the town from end to end,
And half its people, too:
They know where half their patrons live,
What half their patrons do—
And half their patrons also know
The Carmen through and through.

They're ever in the public eye,
Conspicuous in their place;
Their acts are Comp'ny credit marks,
Or Company disgrace—
Be their's the glory in the fact
They mostly credit face.

We trust them when we send them forth
Alone to do their part;
We trust their sense, their honor, and
Their courtesy and art—
For friendships, yea, and troubles, too,
Do on the platform start.

Their uniforms are tokens
Or service cheerfully
Rendered to the public—
They robes of honor be,
For to many of the people
They are the Company.

Published as a "holiday greeting" in 1916, Our Songs of Labor *looked and read like a union tract. The resemblance was hardly accidental.*

During the latter years of my association with you I have, I must confess, taken a much deeper interest in the progress of the EMBA and in the welfare of the families of its members, than I have taken in dividends, injunctions, political campaigns, rate schedules or hostile criticism.

It took me some years to find out that if the body of workmen, my associates in the operation of these vast enterprises, were well taken care of in their every day requirements, dividends, rate schedules, car service and all other things would come pretty nearly taking care of themselves....

Some years ago you were operating under a more or less paternalistic system of management. I admit I participated in that system in the early years of my administration. Then you appeared thankful for the things that were handed out to you. The management took great pride in announcing a voluntary increase of two cents an hour granted to motormen and conductors. We don't have any more of that now. Things have changed. You have the right to ask for increases in wages or shorter hours. You do ask for it, and the increases run from seven to ten or more cents per hour.

But that is only a part of it. You have a much bigger responsibility than you used to have. It took me some years to learn that the way to do business was to share the responsibility with the others who were working on the same job. In the last four or five years the efforts of Brother Way and departmental heads have been devoted to trying to place responsibility on people further down the line.

World War I and Aftermath

The Great War, as it was then called, was the defining event of the 1911-1920 period. Some of its impacts on the TMER&L system — the surge in demand for industrial power, the inflationary pressures on revenue and wage rates — have already been noted, but the conflict affected the utility even more directly. Between 1914 and 1917, as the nation drifted from neutrality to "preparedness" to full-scale involvement, TM moved with it. A total of 609 employees entered the military. Their places were taken, in many cases, by women, who proved as adept at substation work as they had in office jobs. The Cold Spring shops aided the effort by building a wooden destroyer atop one of the system's flatcars; it sailed down the streets of Milwaukee as part of the Navy's recruiting efforts. The system's employees regularly exceeded their quotas in Liberty Loan drives, and large groups gathered in the PSB auditorium to watch films of "stricken Europe." (Most gatherings ended with the singing of *Over There.*) As part of the national effort to conserve food, EMBA members were encouraged to plant gardens on vacant land owned by the utility. The charge was three dollars per quarter-acre plot, plowing and harrowing included, and the system's gardeners showed off their "war food" at a series of "Country Fairs" staged in the PSB auditorium.

The TM system was a patriotic stronghold during World War I, but the war and its aftermath were also a period of severe shortages. James Mortimer and S.B. Way struggled to keep an adequate supply of coal on hand, and the transit system was perennially short-handed. Mortimer and Way also struggled with a shortage of capital. Material costs doubled during the war, and TMER&L needed a steady infusion of cash to meet its current needs and to fund its ongoing expansion. Utility stocks, unfortunately, had fallen out of favor on Wall Street, torpedoed by their own modest returns and an alarming rise in the number of street railway bankruptcies (from 46 in 1915 to 76 in 1919). An internal study done in 1920 noted the "progressive deterioration" of the market for TM's stocks and bonds, and James Fogarty, the

TMER&L's salesmen marched as a unit in a 1916 Preparedness Parade.

World War I opened some non-traditional jobs to women, including this Kenosha streetcar conductor.
(State Historical Society of Wisconsin)

North American Company's secretary, observed that street railway stocks were "in general disrepute."

Unable to attract large national investors, the utility turned to small local investors. In August, 1918, Wisconsin Gas & Electric put $300,000 in 6.5-percent notes on the market. To management's surprise, they were gone within weeks. The sale was so successful that TMER&L entered the field, offering $3.6 million in 7-percent notes before the year's end. The offer attracted 5,500 investors, nearly all of them Wisconsinites and most of them Milwaukeeans, including a large number of TM employees. With the push provided by a new Securities Department, subsequent issues of preferred stock were just as successful. The TM system thus became a national pioneer in what S.B. Way called "home-town financing." The new strategy provided benefits for everyone: a fresh source of capital for the utility (without brokerage fees), a sound investment for local buyers, and a substantial increase in good will. The system's managers were well aware that sharing ownership with their customers could help to blunt some of the constant political criticism. As S.B. Way put it, "A sufficiently wide distribution [of securities] among the railway patrons should form a more logical basis of public ownership than municipal ownership."

Another shortage was not so easily remedied: a shortage of power. After adding a 4,500-kilowatt turbine to the Commerce plant in 1913, Mortimer and Way had done nothing to increase the system's capacity. They relied heavily on hydroelectricity generated by two dams on the Wisconsin River, one at Wisconsin Dells and the other at Prairie du Sac. (Purchased under a long-term contract negotiated by John Beggs in 1909, the hydro power entered the TM system at the interurban terminal in Watertown.) As power consumption soared, both the Commerce plant and the dams were pushed to their limits. In 1916 TMER&L ordered a pair of generating turbines from General Electric, and in 1917 the system's managers picked a place to put them: a sprawling piece of land on the Lake Michigan bluff near St. Francis Seminary, five miles southeast of Milwaukee's downtown. Plans for the aptly named Lakeside Power Plant were announced in April, 1917, with construction scheduled to begin within a few months.

The United States entered World War I within days of the announcement. The conflict put TMER&L in a painful double bind: It raised the demand for industrial power to new heights and, at the same time, created shortages that forced the utility to postpone the power plant that could meet that demand. Work did not resume until 1919, when the power deficit was entering the danger zone.

To complicate matters, TM had committed itself to a technology that was still under development: the use of pulverized coal to generate steam. The new system was the brainchild of John Anderson, a pivotal figure in American central-station technology. Scottish by birth (a sea captain's son) and a marine engineer by training, he was yet another alumnus of North American's St. Louis system. In 1912, after serving a six-year "apprenticeship" in St. Louis, John Anderson came to Milwaukee as TMER&L's chief engineer. Refining an idea from his early

John Anderson, the Scottish-born engineer who would make pulverized coal the utility industry's standard fuel

years aboard coal-fired steamships, Anderson set out to prove that powdered coal, because of its vastly increased surface area, would burn hotter, cleaner, and more efficiently than the lump variety. With the backing of S.B. Way and the able assistance of Fred Dornbrook, a veteran power plant engineer, he turned the East Wells (Oneida Street) powerhouse into a test facility. Grinding mills were installed, dryers and injectors were fabricated, and boilers were redesigned.

The work began as early as 1914, but the critical experiments were conducted in 1919. In the first tests, pulverized coal produced enough heat to melt the furnace bricks and trip the safety valves on the steam pipes. "It was just a trifle nerve-racking," recalled

Anderson, "to note one failure after another." By the middle of November, 1919, after an endless series of design changes, the Scotsman had proved that his system could generate as much steam as a conventional furnace with less than half the coal. He had also removed, in a single stroke, practically all constraints on boiler size. Because the coal burned in suspension rather than over a fixed grate, boiler chambers could be as big as barns.

Pressed for capacity and pressed for time, Mortimer and Way decided to use Anderson's prototype as the technical foundation for the Lakeside plant — a staggering leap of faith. Lakeside was conceived as the central power plant for the entire TM system, and a failure there would have been nothing short of disastrous. "One must admire," wrote the *Electric Railway Journal,* "the courage displayed in daring to pioneer on so big an undertaking as a 200,000-kilowatt station completely built around the idea of burning coal in pulverized form." Anderson never doubted his system. As later events would demonstrate, pulverized coal was the single most important advancement in steam generation until the advent of nuclear power in the 1950s.

Putting the system to work was another matter, for the Lakeside plant was built at the worst possible time. When excavation began in February, 1920 (only weeks after Anderson's final experiments), the soil was frozen so hard that the "ground-breaking" was accomplished with dynamite. Local workers were in such short supply that TM was forced to open a semi-permanent labor camp on the site; it was soon filled with laborers from throughout the Midwest. Cement and steel were both scarce, and a rail strike delayed shipments of any available supplies. Material costs spiraled upward with inflation, hampering TM's ability to finance the project. Construction crews pushed on despite the obstacles. The power shortage had reached critical proportions, and Lakeside was the only light at the end of the tunnel.

James Mortimer was not on hand to see that light come into view. After 10 years at TMER&L and nine at the helm, Mortimer suffered a power shortage of his own. Although he presided over the entire North American system, the young executive was a hired manager, not a major stockholder, and the death of James Campbell in 1914 had left him in a somewhat vulnerable position. As long as North American's stock was widely dispersed, Mortimer had no worries but, in the early months of 1920, another tycoon, Harrison Williams, formed a syndicate to buy the Campbell family's stock. He promptly proceeded to take over the company.

Born in Ohio and based in New York City, Williams already controlled a

Faced with a severe power shortage, TMER&L began construction of its Lakeside Power Plant in 1920. Milwaukee workers were so scarce that the company had to open a resident labor camp (right) on the site.

sizable utility empire. Using the earnings from a bicycle factory and a tire business as his grubstake, the entrepreneur had formed American Gas and Electric in 1906 and Central States Electric in 1912. His holding companies controlled utilities from Cleveland to the Gulf Coast, and Williams planned even greater things for North American. Like the new owner of a major-league baseball team, his first step was to fire the manager. On April 27, 1920, James Mortimer's association with North American and TMER&L came to an end. He was 40 years old.

Mortimer's tenure at North American was, it turned out, the high point of his management career. He took positions with two distressed street railway companies, but the one-time whiz kid soon entered the quieter field of investment securities, which remained his specialty until his death in 1950. Although his term with TMER&L was relatively brief, James Mortimer left an indelible mark on the system. Working with and through S.B. Way and other executives, he turned a traction company into an electric company. Working with and through employees on all levels, he created a corporate culture that endured for decades; the EMBA was a genuine breakthrough, and James Mortimer was its founder. He was also a good steward. Despite stagnation on the transit side and constant commotion on the political side, the system's gross income tripled during his tenure, and its net income doubled. Mortimer's achievements were dramatic, but they were also a prelude. By the time he stepped down in 1920, the TMER&L system was perched on the threshold of the most explosive growth in its history.

Harrison Williams, a capitalist who played to win, took control of North American in 1920.

Within weeks of Williams' rise to power, the EMBA was hosting a farewell banquet for James Mortimer.

Chapter 5

1920-1929

Superpower

It was still possible to dream of empire in 1920. For all the dramatic progress of the previous decade, there were still new markets to serve, new opportunities to seize, new worlds to conquer. In a period of roaring prosperity, dreams of empire spawned gigantic holding companies, and a new term entered the nation's vocabulary: "superpower." It meant, first of all, interconnected utility systems that spanned states and even regions. Superpower also meant a large step forward in the evolution of the business. In the late 1800s, utilities had been points of light in the American darkness, extending only to the nearest city limits. In the 1920s, they reached nearly everyone. Neighboring systems were locked in a frantic struggle for territory, and an "electric blanket" practically covered the country by the decade's end.

The TMER&L system shared fully in the expansionism of the times. The 1920s witnessed other developments — a subtle shift in the company's public image and political fortunes, novel solutions to old problems on the transit side — but the decade's keynote was growth. Under two presidents, both company veterans, the utility edged steadily outward from its original strongholds, lapping against the territories of its competitors. The system's growth was relentless, almost reckless, and it produced an empire that stretched from Michigan on the north to Illinois on the south. Headlong growth was accompanied by a groping toward maturity. By the time the sober realities of the Thirties dawned, the Wisconsin Electric system had emerged in substantially its modern form.

The Patriarch's Return

To many Americans, Harrison Williams was superpower incarnate. Soon after he took control of North American in 1920, Williams made it one of the fastest-growing holding companies in the nation. The properties he had inherited — utilities in Milwaukee and St. Louis, a Kentucky coal subsidiary — became regional capitals of an empire that sprawled practically from coast to coast. In 1922 Williams brought the Cleveland electric utility into his corporate family. In 1925 he purchased Western Power Company, which served San Francisco and California's Central Valley. In 1928, after years of persistent effort, Williams acquired Potomac Electric Power Company, the utility for Washington, D.C., and adjacent communities in Maryland and Virginia. By 1929 North American controlled 7 percent of the

"Electricity turns the wheels" — an illustration from TMER&L's 1926 annual report

nation's power supply, serving more than 1 million customers in 10 states. Profits rose accordingly. After resting comfortably at the $3 million to $4 million level during the Mortimer years, the holding company's annual earned surplus rocketed to $12.2 million in 1927 and $25.2 million in 1929.

(Williams did not restrict his activities to the utility field. In 1922 North American launched an imaginative subsidiary known as Wired Radio, Inc. It was, in essence, a closed-circuit radio network, a subscription service that used electric lines to bring musical "super-programs" into office buildings, stores, and even private homes. The company is better known today as Muzak, Inc.)

The North American empire assumed the characteristic shape of all major holding companies: the pyramid. At its base were dozens of operating subsidiaries, including TMER&L and WG&E. An assortment of subholding companies occupied the middle layers, and at the apex was the aptly named New Empire Corporation. The key to the entire structure was the time-honored principle of leverage. Like his counterparts in other companies, Williams recruited tens of thousands of outside investors, retaining just enough stock to ensure control. In 1930 he stood atop a pyramid of 91 corporations with a combined value of $894 million, and he controlled the whole system with a stock interest of $86 million.

Although he made the deals and set the policies, Harrison Williams did not manage the system single-handedly. Operating details were left to a Manhattan staff headed by Frank Dame, who had left General Electric to work for Williams in 1912. Dame relied heavily on a Milwaukee boy and TM alumnus, Edwin Gruhl. A 1908 graduate of the University of Wisconsin, Gruhl had spent four years on the Railroad Commission's staff before joining TMER&L. He quickly became James Mortimer's right-hand man, following his boss to New York in 1914 and surviving his ouster in 1920. Only months after Mortimer's resignation, Edwin Gruhl became North American's general manager, second only to Frank Dame in the operational hierarchy. Although he was responsible for all North American subsidiaries, Gruhl maintained especially close ties with the Milwaukee system. In 1924 he married S.B. Way's daughter, Helen, making him one of the few executives in America who supervised his own father-in-law.

While Dame and Gruhl oversaw the growth of his empire, Harrison Williams lived the life of an uncrowned king. He owned five homes, including a 30-

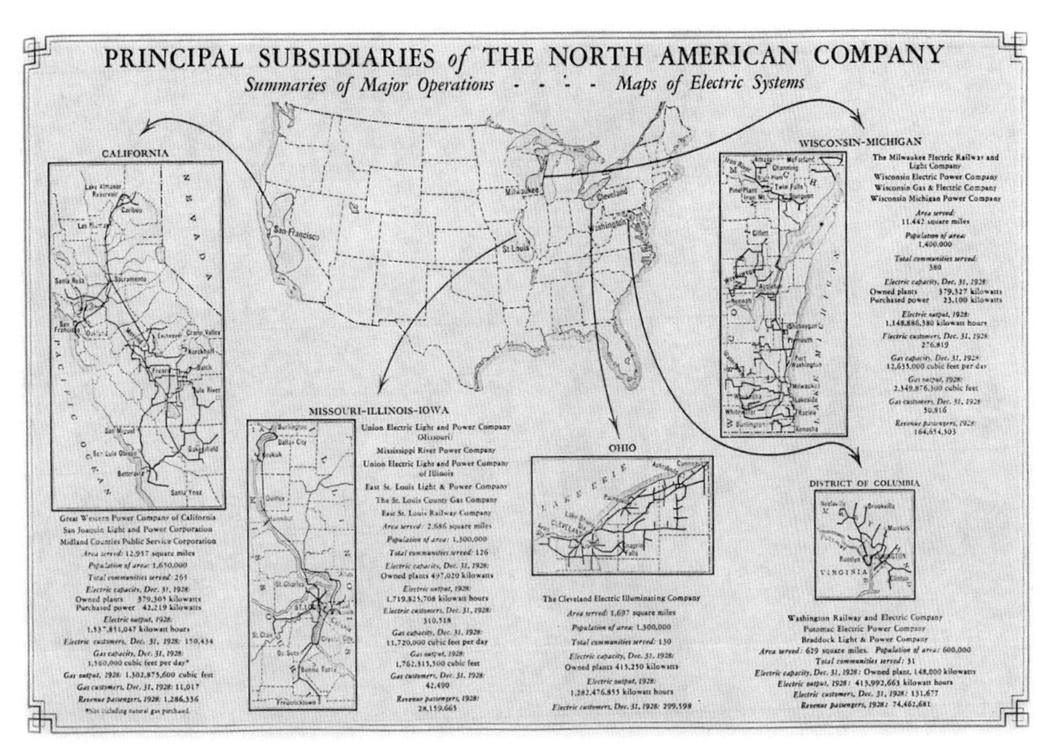

With subsidiaries scattered from California to Maryland, the North American Company was very nearly a coast-to-coast concern in 1929.

Edwin Gruhl, the Milwaukee native and TM alumnus who became North American's general manager in 1921

room mansion on Park Avenue, a villa on Capri, a townhouse in Paris, a mansion in Palm Beach, and an estate on Long Island. (Williams commuted to Manhattan from his Long Island home by hydroplane.) He was an art collector of wide renown, and his yacht, the 400-foot *Warrior,* was said to be the world's largest. In 1926, after his first wife's death, Williams married Mona Bush, a Kentucky belle who had once been the wife of Henry Schlesinger, a socially prominent Milwaukeean. Mona, who ultimately went through five husbands, became "the best-dressed woman in America" during her years with Williams. If the Twenties were a period of reckless extravagance for America's upper crust, Harrison and Mona Williams practically defined the style.

The company that funded Mona's shopping sprees changed rapidly in size and scope, but Harrison Williams did not tinker with its basic character. The fluid, collegial style adopted by Henry Villard still applied at North American, and the subsidiaries continued to operate as a federation of semi-autonomous enterprises. Few people at TMER&L noticed the change in ownership. There were no purges; the only change, in fact, took place at the very top. On May 25, 1920, less than a month after James Mortimer's departure, the TM board elected a new president: John I. Beggs. The patriarch had returned. The company's directors viewed the appointment as temporary. They noted that TM needed "an independent, active man who shall devote all his time" to the business, but that "at the present time such a man is not available." (Beggs was 72 in 1920, and his outside interests were legion.) TM's old leader was supposed to serve only until a permanent successor had been named, and the directors placed him on a search committee to find such a man. As an important investor in North American (with a 2.5-percent interest in 1920), Beggs could afford to take his time. He may have looked high and low but, characteristically, the "interim" executive could find no one worthy of inheriting his mantle. Beggs held on to the TM presidency until his death.

It would be an exaggeration to state that John Beggs had been waiting in the wings for a chance to reclaim his rightful throne. In the nine years since his resignation, he had become a full-time tycoon, managing some companies, investing in others, and generally multiplying his millions. Beggs had returned to Milwaukee from St. Louis in 1916, taking a small apartment in the Yankee Hill section north of downtown. City directories of the time list him as president of Wisconsin Traction, Light, Heat & Power, the Appleton utility he had taken over in 1900, but that was only the tip of the iceberg. Beggs had interests on a par with a Wall Street banker's, and he continued to build his portfolio after rejoining TMER&L. By 1925 he was a one-man interlocking directorate, serving on the boards of no fewer than 53 corporations. Some were regional; Beggs was board chairman of J.I. Case, the Racine-based farm equipment giant, and he was a dominant force on the First Wisconsin Bank board. His other interests were spread across the country; Beggs was a principal in ventures that included an Arkansas railroad, a Louisiana paper mill, an Atlantic City hotel, and a tract of 27,000 acres on Florida's Atlantic coast. Work, obviously, remained the capitalist's

The lion in winter: John Beggs was 72 years old when he resumed the TMER&L presidency in 1920.

reason for being. "A good constitution," he said in 1920, "a love for work, and the ability to get along with four or five hours sleep has enabled me to accomplish much work." Beggs was so busy on so many fronts that his secretaries carefully divided his days into 15-minute segments.

The presidency of TMER&L was, therefore, a part-time job for John I. Beggs. He apparently never took an office in the Public Service Building, preferring to work out of the Beggs Investment Company's quarters above the First Wisconsin Bank. A 1922 newspaper report described the TM system as "his business hobby." It was, however, a hobby that Beggs took seriously. He had retained close ties to the company since 1911, playing an active role on the board and

The EMBA continued to be an organization for all seasons, sponsoring hockey teams in winter ...

participating faithfully in Veterans Association affairs. Beggs knew the system intimately and, of all his various business interests, TM was probably closest to his heart.

John Beggs worked part-time, but even a part-time president might have cleaned house and reordered the system's priorities. Beggs had no such inclinations. Although he established several new policies, the patriarch made no attempt to shift the balance of power back to the traction side. He did not reinstitute the system of company spies, nor did he tinker with the existing management structure. Beggs could read a financial statement as well as anyone, and he found no reason to change a successful operation. In an interview with the *Milwaukee Journal* (May 27, 1920), he downplayed the significance of his return:

> *I think I can safely say there will be no important changes in the organization of the company. There is no reason why changes should be made. It is simply the old commander coming back to the old guard. Practically every man over there who occupies a place of importance was trained by me, and trained from the ranks up.*

Beggs' reference to "the old guard" might have surprised S.B. Way, John Anderson, and other key recruits of the Mortimer era. It was Way, continuing as general manager, who kept the system running smoothly while the president managed his other interests. But Beggs had clearly signaled his intent to leave well enough alone. The transition to a new administration was seamless; efforts of the Mortimer years were continued and even expanded.

Beggs' approach to the company's welfare programs was typical of the style that characterized his second term. Employee relations had never been his forte, and a younger Beggs would almost certainly have considered the programs an extravagance. He

... and bathing beauty contests in summer.

did nothing, however, to curb their growth, and the Employes' Mutual Benefit Association spread its umbrella even wider during his tenure. The number of house and office calls made by EMBA doctors soared from roughly 10,000 a year before 1920 to 41,491 in 1924. Low-cost dental care was added to the plan in 1921, and a new "health service" was unveiled four years later. In an effort to help employees "increase their enjoyment of health, avoid possibility of sickness, and prolong life," EMBA doctors offered free annual checkups to all members. Terms like "wellness" and "holistic health" hadn't been invented yet, but the service clearly anticipated modern trends. On other fronts, a bathing beauty contest was added to the company picnic in 1922, and an informal group called the Non-Golfers' Protective Association was organized in about 1925. (Its sole purpose was "golf prevention," and its proposals included "the ploughing up of every golf course in the land and the planting of popcorn.") As the economy boomed, nearly 90 percent of the system's employees entrusted their savings to the Employes' Mutual Saving, Building & Loan Association; by 1927 80 percent of the system's full-time, married employees owned their own homes.

Employee welfare programs required nothing but Beggs' tacit approval, but he was forced to act more decisively on another matter: finishing the Lakeside Power Plant. The plant had been a work in progress

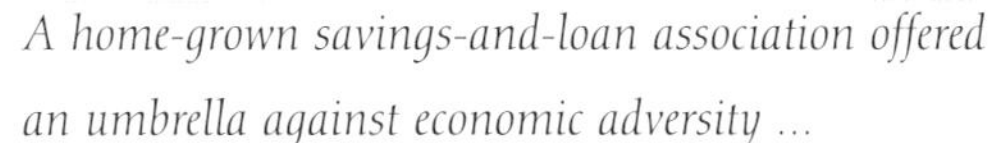

A home-grown savings-and-loan association offered an umbrella against economic adversity ...

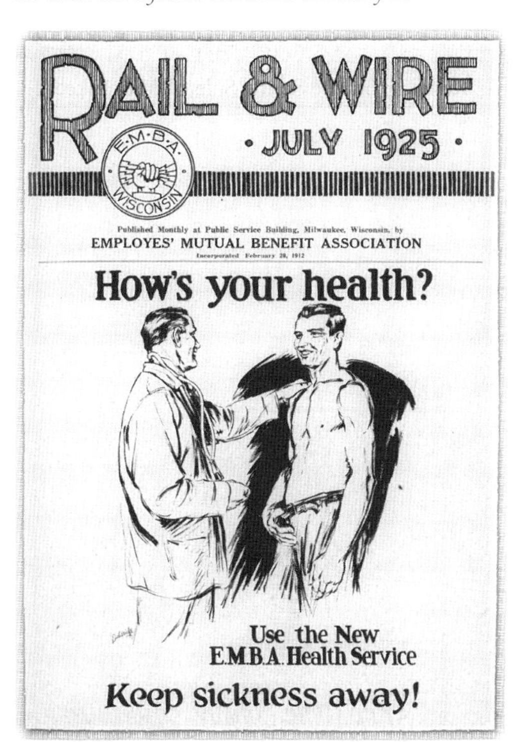

... while the EMBA medical program kept employees in top physical health.

The Lakeside Power Plant was built with extraordinary speed. These views document the progress made in just seven months, from September, 1920, to April, 1921.

when James Mortimer left, and it became the first item on Beggs' agenda. North American's new managers had halted construction in April, 1920, "pending determination as to the actual necessity of its completion this year." Beggs put the crews back to work immediately. Demand for power was rising steadily and, by the autumn of 1920, the system faced the most serious power shortage in its entire history. Equipment in existing plants was overloaded by as much as 50 percent. Milwaukee's electric foundries were forced to rotate weekly shutdowns. Residential customers were begged to cut their consumption by at least 20 percent. Desperate for power, TM purchased the surplus output of Allis-Chalmers, Pfister & Vogel, and other large industries with independent plants. The Lakeside crews, in the meantime, were working at a breakneck pace, and they finished the plant in record time. The new turbines were turned over for the first time on December 15 — only 10 months after ground was broken. Following a few months of testing and troubleshooting, the plant was dedicated on April 15, 1921, "to the service of mankind."

Lakeside was a success from the beginning. The plant began to set world records for efficiency in 1922, and John Anderson's team added features that steadily widened its lead. High-pressure boilers, producing steam at 1,200 pounds per square inch, were installed in 1926, reducing coal consumption significantly. New turbine sets were added every year or two; the plant's capacity soared from 40,000 kilowatts in 1921 to 160,000 in 1926 and 310,800 in 1930.

The tube assembly for one of Lakeside's boilers dwarfed the men working on it.

New boilers and turbines were added every year or two. Dignitaries at a 1926 dedication included John Anderson and his daughter Miriam, an engineering student at the University of Wisconsin.

Even before reaching its final capacity in 1930, Lakeside was the linchpin of the entire TMER&L system ...

(Milwaukee Public Library)

... and the world center of pulverized-coal technology.

The turbine room at Lakeside was superpower personified.

Lakeside became a regional giant, using twice as much water daily as the entire city of Milwaukee, and it served as the workhorse for a sprawling utility. By 1925 the plant generated 86 percent of the TM system's electricity, leaving the Commerce and East Wells facilities to produce steam heat and standby power. As the prototype of a technology that would quickly win universal acceptance, Lakeside was also the powerhouse that made Milwaukee famous. John Anderson spread the gospel of pulverized coal far and wide, and he attracted scores of visitors who soon became disciples. To utility engineers around the world, Milwaukee meant Lakeside, not beer.

Financing the plant had required a little ingenuity. Like most American utilities, the TM system was neck-deep in debt. In 1920 TMER&L's debt was 68 percent of its total capitalization — somewhat better than the industry average of 79 percent. Understandably, bankers were reluctant to gamble on an expensive new project with so many creditors ahead of them. North American solved the problem by creating a new subsidiary: the Wisconsin Electric Power Company. WEPCO's sole asset was the Lakeside Power Plant, which TMER&L agreed to lease and operate until 1956 (later 1970). Freed from TM's debt load and interest limitations, WEPCO attracted more than enough capital to finish the plant.

Much of the capital was local. Beggs continued and expanded the Mortimer-Way emphasis on "home-town financing," using newspapers, billboards, and even streetcars to advertise new issues of preferred stock. By 1925 roughly 25,000 Wisconsinites had invested in the system: 16,000 in TMER&L, 5,000 in Wisconsin Electric, and 4,000 in Wisconsin Gas & Electric. Their total stake exceeded $30 million. The confidence of local investors, combined with John Beggs' ever-frugal management, broadened the lead company's margin of safety. TMER&L's debt fell from 68 percent of its capitalization in 1920 to 65 percent in 1925 and 55 percent in 1926. "Frenzied finance" may have ruled the American economy in the Twenties, but John Beggs, typically, sat out the stampede.

In 1922 TMER&L purchased the Milwaukee Northern Railway, whose assets included a small power plant in Port Washington.

Finishing the Puzzle

Opening Lakeside was like breaking a logjam. After two years of embarrassing power shortages, the TM system had more than enough capacity, freeing Beggs and company to take on the central task of the Twenties: expanding the TM empire. It was a task well-suited to the times, well-suited to the imperial tendencies of Harrison Williams, and certainly well-suited to John I. Beggs. Beggs had long envisioned Milwaukee as the capital of a utility empire covering eastern Wisconsin, and his interurban railways had formed the partial outline of that empire. After an absence of nine years, Beggs returned to the work of finishing the puzzle.

The first piece he added was, to no one's surprise, another interurban railway. In July, 1922, after months of negotiations conducted by Beggs himself, TMER&L purchased the Milwaukee Northern Railway for $1 million. Completed in 1908 by Detroit capitalists, the line had preempted the corridor between Milwaukee and Sheboygan. In Beggs' view, the purchase restored a natural order; the Northern took its place as one more spoke in the wheel of TM interurbans radiating outward from Milwaukee. The acquisition included 57 miles of track, a small powerhouse in Port Washington, and a fleet of perhaps 20 cars, which were promptly upgraded. (TM's old funeral car, an elegant steel-wheeled hearse, was rebuilt as a parlor car for the Northern in 1923.) The line operated independently until 1928, when it became part of the regular TM system. In addition to rail service, the Northern provided one more route for the spread of electrical service into the hinterland.

Beggs' next step was a good deal larger and considerably easier. Since 1900 he had been the principal owner of Wisconsin Traction, Light, Heat & Power, the utility that provided electric, gas, and rail service in Appleton and nearby communities. Its largest rail holding was an interurban connecting Kaukauna, Appleton, and Neenah. In May, 1923, Beggs sold WTLH&P to North American, enlarging his own fortune and giving the holding company another foothold in eastern Wisconsin. The Appleton powerhouse

The Appleton utility was added in 1923. Built astride the Fox River, its powerhouse was both a steam plant and a hydroelectric facility.

The Peninsular Power Company joined the system in 1924. Hydro plants like this one on the Brule River expanded the TMER&L empire into Upper Michigan.

was promptly converted to the pulverized-fuel system, and in 1928 a new gas plant was erected to serve customers in Appleton, Neenah, and Menasha.

Serendipity played a role in the third major acquisition. In 1911 a group of investors led by Daniel Mead, a University of Wisconsin engineering professor, had formed the Peninsular Power Company to serve a cluster of iron mines in Michigan's Upper Peninsula. Under Mead's direction, the company built a hydroelectric dam at Twin Falls on the Menominee River, the border between Wisconsin and Michigan. Mines in the Menominee range, including the shafts at Iron Mountain, quickly converted to electricity, and Mead and company built a second dam on the Brule River in 1917.

Peninsular Power was a wholesale operation; its only customers were mines. In the boom following World War I, the investors (who included several UW professors) had two choices: to become a full-fledged utility, providing retail power to anyone who wanted it, or to sell out. Unwilling to expand on their own, they chose the second option. Their first buyer, however, backed out of a tentative deal, arousing the interest of Harrison Williams and John Beggs. Even Beggs had never dreamed of reaching as far north as the Upper Peninsula, but the opportunity was irresistible. In December, 1924, the North American Company purchased Peninsular Power, making the TM system a two-state utility.

By 1925 North American had three strategic bases in the region: Milwaukee, Appleton, and Iron Mountain. The next logical step was to grow from all three until their service areas met, creating one cohesive superpower system. Substantial progress was made toward that goal. Peninsular Power grew both north and south, reaching paper mills and municipalities on both sides of the state line. WTLH&P concentrated its efforts on the area north of Appleton; by 1925 the utility served 10,415 customers in seven counties. The system's greatest growth took place on the outer fringes of Milwaukee. As TMER&L's regional arm, Wisconsin Gas & Electric pushed north and west of the city, straining toward a connection with Appleton. Dozens of small plants were acquired and retired, their output replaced by cheaper power from the Lakeside plant. (In 1925, for instance, Kewaskum residents said goodbye to the little light plant in Rosenheimer's grain elevator.) In the decade following 1920, the number of communities served by WG&E increased from 39 to more than 100, including Whitewater, Waukesha, Fort Atkinson, Hartford, West Bend, and Grafton.

In the race for territory, the TMER&L system had always been ahead of the pack. As the first utility formed by the nation's first utility holding company, it had achieved relative maturity long before other systems coalesced. By 1920, however, the field was anything but clear. To the north and west of Milwaukee, Beggs and his associates faced intense competition from Samuel Insull, already a legendary figure in the utility industry. Based in Chicago, his Middle West Utilities Company practically defined superpower; by 1926 it sprawled across 32 states and controlled 12.5 percent of the nation's power supply. (North American was second or third to Middle West in most years.) Insull had not overlooked opportunities in Wisconsin. In 1917 Middle West Utilities purchased the Wisconsin River dams that furnished TMER&L with much of its power. With the completion of Lakeside in 1921, the need for that hydro power evaporated. Looking ahead, Insull's local managers had already begun to reach a broader market. They expanded aggressively throughout south-central Wisconsin, ultimately reaching the Illinois border on the south and WG&E territory on the east.

The dams were soon overloaded, and the Insull interests began to look for a plant site on Lake Michigan, which offered an endless supply of fresh water as well as access to the

The race for territory often went to the first company that could string its wires into an unserved area. This erecting crew had to ford a river west of Mukwonago in 1924.

Great Lakes coal fleet. In 1922 Middle West Utilities beat WG&E to the punch by purchasing a utility that served a narrow corridor between Fond du Lac and Sheboygan. Perhaps its most important asset was a steam plant in Sheboygan. The purchase led, in 1924, to the formation of a company that combined all the Insull interests in the southern portion of the state: Wisconsin Power & Light. Thus began one of the state's leading utilities, and thus ended TM's hopes of an unbroken connection with its sister in Appleton.

Hemmed in by Wisconsin Power & Light to the north and west, John Beggs looked farther north to finish the TM empire. The dominant utility in northeastern Wisconsin was (and is) the Wisconsin Public Service Corporation, which served a Y-shaped area stretching north from Sheboygan along both sides of Green Bay, reaching Upper Michigan on one side of the bay and Door County on the other. WPS controlled practically all the territory that separated North American's Appleton utility from its sisters in Milwaukee and the Upper Peninsula. The company's guiding light was Clement Smith, a canny investor who had won early renown as a traction engineer. Although the city of Green Bay was the WPS system's operational center, Smith kept its corporate headquarters in Milwaukee.

John Beggs had reason to believe that Wisconsin Public Service would, in time, join the North American empire. Both he and George Miller, another TMER&L director, were major investors in the company, and both men had seats on the WPS board. Miller was the head of Miller, Mack & Fairchild (now Foley & Lardner), the law firm that served as TM's general counsel. Clement Smith was also Miller's brother-in-law and one of Beggs' best friends; in 1922 he gave the patriarch a solid gold serving plate for his seventy-fifth birthday, which was lavishly celebrated on Beggs Isle. Although they were locked in a spirited struggle for territory, it was expected that WPS and North American would some day combine their systems on terms agreeable to both.

The gentleman's agreement unraveled in 1925. When Peninsular Power was forced to cut off some customers during a period of low water, Beggs and Way decided to connect their northern satellite with the Lakeside Power Plant. The connection would enable Peninsular to send hydro power south during the spring run-off and to use Lakeside's output during dry spells. The system's agents immediately began to buy right-of-way for a 132,000-volt line linking Milwaukee, Appleton, and the dam at Twin Falls. The problem, of course, was that the new superpower line had to pass through WPS territory, and it was projected to cross the Fox River at Green Bay, the nerve center of the WPS system. Clement Smith took exception almost instantly. Convinced that North American was invading his

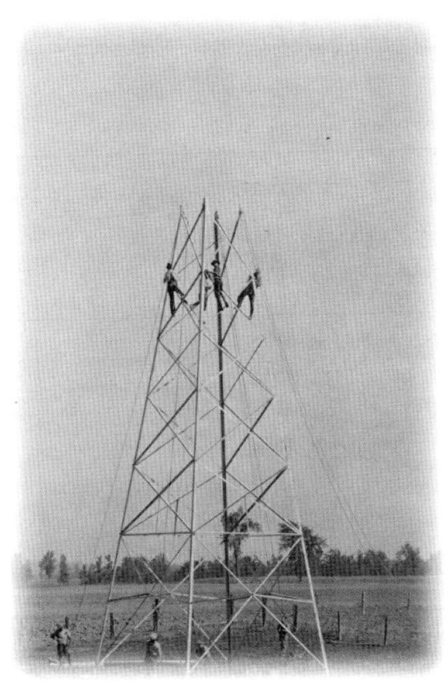
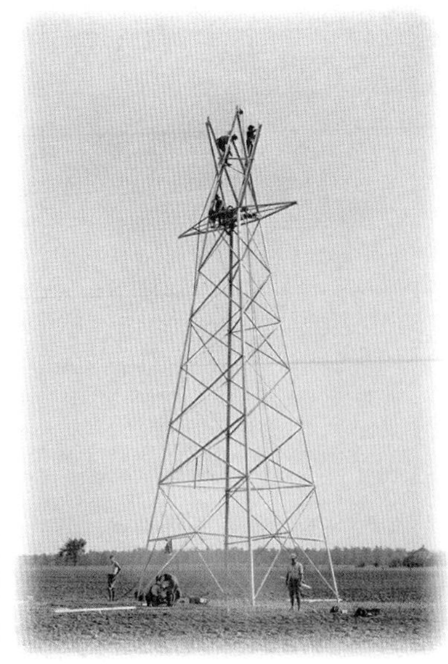
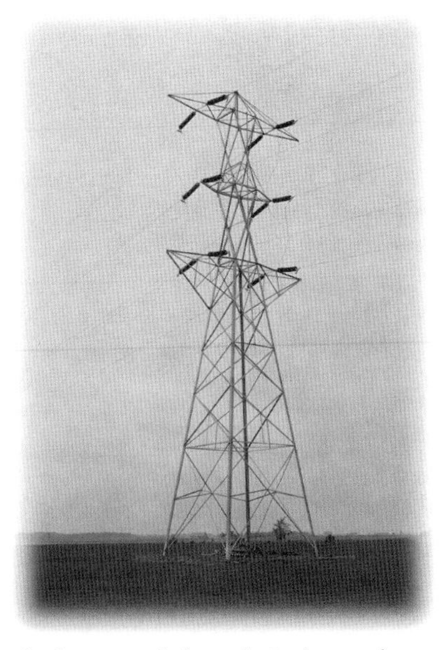

In 1925 North American erected a 132,000-volt transmission line to link its holdings in Milwaukee and Upper Michigan. The line passed through the heart of Wisconsin Public Service territory, sparking a pitched legal battle.

empire with intent to conquer, Smith tried to block the intruders by purchasing narrow strips of land at right angles to the planned corridor.

In the legal battle that followed, George Miller sided with his brother-in-law, which caused, among other things, the termination of his firm's contract as TM's general counsel. WPS lost the case, and construction of the line proceeded. Although Smith was not known as an impetuous man, he quickly decided to dispose of his system rather than come to terms with North American. In July, 1925, he sold WPS to Standard Gas & Electric, a Minnesota-based holding company organized by H.M. Byllesby. The Byllesby interests showed no desire to accommodate North American, and the battlefront was frozen. From Beggs' point of view, a golden opportunity had vanished into the thin air over northeastern Wisconsin. (The companies' paths finally converged in 1995, when the Milwaukee utility announced plans to merge with the old Byllesby group's flagship: Northern States Power Company.)

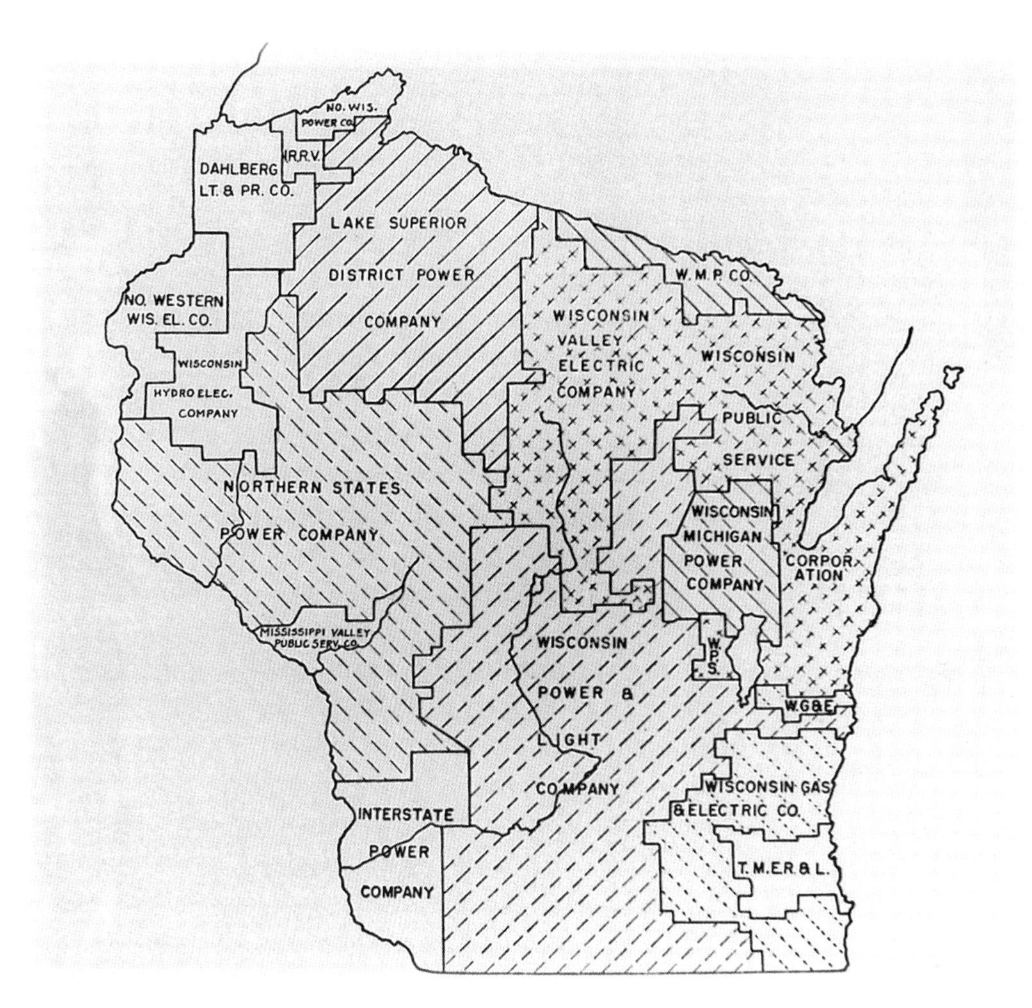

By the mid-1920s, the race for territory had made Wisconsin a jigsaw puzzle of electric service areas that made little technical or administrative sense. (from Forrest McDonald's *Let There Be Light*)

The friction between Clement Smith and North American was entirely typical of the 1920s. The utility wars amounted to a high-stakes game of Monopoly, each company trying to buy and improve property before its competitors could act. Wisconsin became a hodgepodge of utility systems, both large and small, that were constantly changing size, position, and ownership. The overall trend was toward consolidation. As holding companies acquired their smaller competitors (generally at a premium), the struggle for territory became a battle of giants. Competing systems nibbled at each other's flanks, threw roadblocks in each other's paths, and frequently engaged in guerrilla warfare. Some border disputes were settled by friendly negotiation. Others were "settled" by keeping line crews on the job seven days a week, preempting territory that other companies coveted. The so-called "zone of darkness" retreated all the while; the beneficent by-product of the struggle was power and light for nearly everyone.

Wisconsin's utilities, to use a metaphor from high-school science, were like crystals growing in a laboratory dish, edging out from specific nuclei until they met at their margins, creating an intricate, interlaced pattern that was more the product of chance than design. There would be some jostling at the margins for years to come, but the solution had crystallized by the mid-1920s. The lines were drawn; the puzzle was finished. Wisconsin was divided into large blocks that rivaled in complexity the most flagrantly gerrymandered political districts. North American companies served Elkhart Lake but not Plymouth, Hartford but not Hustisford, Appleton

John Beggs died in 1925, as the utility wars were winding down. This 1922 photo, labeled "The Old Guard" in Beggs' own hand, shows him flanked by long-time TMER&L directors Fred Vogel (left) and Charles Pfister.

but not Kaukauna. The service areas defied logic in some cases, but they have endured to the present. Wisconsin Electric is serving essentially the same communities that were part of the system in 1925.

John Beggs died as the puzzle was nearing completion. On October 17, 1925, just two days after his last trip to the office, the patriarch passed away in a Milwaukee hospital. He was, at 78, the last of the Edison-era pioneers still active in the business, and he died, as he had wished, with his boots on. Although he ruled with a firmer hand in his first term, Beggs had presided over the TM system during a period of tumultuous, dramatic growth, and the company sent its "old commander" off in grand style. His only Milwaukee service was in the auditorium of the Public Service Building, perhaps his most durable monument. An EMBA team conducted the ceremony with all its customary pomp, and some of Beggs' "boys," now middle-aged or older, gave heartfelt tributes. At the time his coffin was carried to the Milwaukee Road station, every streetcar in the city came to a complete stop, and every motorman and conductor removed his cap. Beggs was finally interred at Grace Methodist Church in Harrisburg, where his utility career had begun 42 years earlier. That career had been both productive and profitable. John Beggs was reputedly the richest man in Wisconsin at his death; his estate, which became the subject of a prolonged probate struggle, was conservatively valued at $40 million — more than $350 million in 1996 dollars.

The TM system missed not a beat when its leader died. Sylvester B. Way, who had been to some degree the power behind the throne for years, became TMER&L's president on October 23, 1925, only six days after Beggs' death. In announcing the promotion, *Rail & Wire* described Way as a throwback to the early John I. Beggs:

> *He had no money, no influence nor powerful friends, no rich or influential relatives to assist him. He grasped the opportunities as he saw them. He modestly attributes his success to diligence, confidence and sticking to one job rather than jumping around.*

S. B. Way, the system's day-to-day head since 1914, immediately succeeded Beggs as president.

The system's new president was routinely on the job before dawn, and he worked into the wee hours when deadlines loomed. Although he was a stern taskmaster, Way demanded no more of his staff than he demanded of himself. "He never asked you to do the job without himself participating in it," recalled one associate. "Hours weren't the only answer, nor was the pay."

Samuel Insull and Clement Smith had smothered North American's hopes of a continuous utility empire across eastern Wisconsin, but the indefatigable Way continued the superpower program. In 1926 TMER&L and the Insull subsidiary in northern Illinois agreed to link their systems at the state line. Designed for use "only in cases of emergency," the connection enhanced the reliability of both systems and reduced the need for standby equipment. The wisdom of the decision was demonstrated in October, 1927, when an equipment failure at Lakeside caused widespread outages; the Insull plant in Waukegan restored TM's power within minutes. The hookup also reflected the growth of a national power grid; Wisconsin became the latest link in a chain of interconnected utilities that stretched from Boston to Upper Michigan.

Although the opportunities were limited, S.B. Way continued to enlarge his system's service area, particularly in the sparsely settled sections of northern Wisconsin and the Upper Peninsula. In 1927 TMER&L's northern sisters, Peninsular Power and Wisconsin Traction, Light, Heat & Power, were combined in a single entity: the Wisconsin Michigan Power Company. The merged operations were more than 50 miles apart, forever separated by Wisconsin Public Service holdings, but the new structure simplified management on the northern edge of TM territory. Headquartered in Appleton, Wisconsin Michigan Power began to grow from a base of 150,000 people in 76 communities.

The TM system that emerged in the Twenties was a study in geographic contrasts. It sprawled from the cutover pineries and hardscrabble mining towns of Upper Michigan to the bucolic farms and densely settled cities of southeastern Wisconsin. Other systems in the state covered larger areas, but none had the diversity, or the dominance, of the Milwaukee-based utility. By 1929 North American's Wisconsin subsidiaries accounted for 52 percent of the state's power capacity and 65 percent of its consumption.

Developing the Territory

Geographic expansion was the dominant theme of the 1920s, but the whole point of expansion was a larger electrical load. It was not enough, even as a defensive measure, to annex territory and hold it with a bare minimum of service. During the Twenties, therefore, the TM system grew deeper as it grew wider; service intensified as the service area expanded. Load-building had been S.B. Way's specialty since his arrival at TM in 1911. As general manager and then president, he continued the campaign on a larger scale than ever before.

Wisconsin Michigan Power added customers daily, but the greatest growth took place in the Milwaukee area. Wisconsin became a predominantly urban state in the 1920s (with city-dwellers rising from 47 to 53 percent of the population), and Milwaukee, in particular, was booming. The city's population swelled from 457,147 to 578,249, the greatest absolute increase in its entire history. Nearby communities, including a variety of new suburbs, were showing equally impressive gains. More people meant more homes, and more homes meant more TMER&L customers. With the Lakeside plant producing cheap and abundant power, electric rates dropped 17 times between 1921 and 1929, putting power within reach of all but the poorest families. The proportion of electric homes in what was called, even then, "Greater Milwaukee" soared from perhaps 35 percent in 1920 to 67 percent in 1923, 83 percent

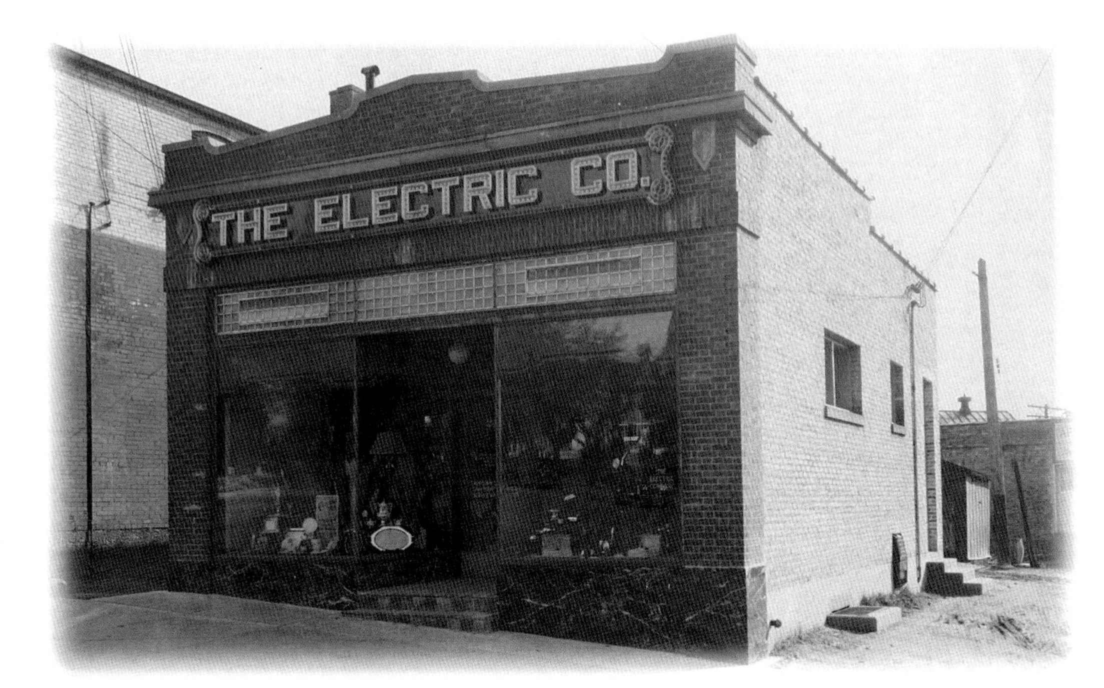

in 1925, and 97 percent in 1927. After years as a rich person's novelty, electricity became nearly universal.

As in the previous decade, it was appliances, not lighting, that made the residential load profitable. TM salesmen redoubled their efforts, traveling door-to-door to pitch washers, refrigerators, and vacuum cleaners. Sales offices in Wauwatosa, West Allis, and North Milwaukee expanded their wares, and the main store in the Public Service Building was enlarged more than once. In 1927 S.B. Way launched the Home Service Bureau to provide "sympathetic counsel and help for the home maker." Its showcase was the new "Electrical Home" in the PSB base-

Consumers could choose from a dazzling variety of lights and appliances on display in the Public Service Building's retail store or (top) in local outlets like the Wauwatosa sales office.

The "Electrical Home" in the Public Service Building's basement showcased the very latest in modern conveniences. Note the industrial-size mixer on the counter and the dishwasher on the sink.

ment, which illustrated "every practical application of electricity" from the laundry to the living room. Classes in baking, cooking, and even lampshade-making attracted thousands of housewives, and Home Service Bureau cookbooks, prepared by "women of wide, practical experience," won enormous popularity. The net result of all the promotions was a steady increase in appliance sales. By 1927, when the PSB store alone generated more than $1.5 million in revenue, 45 percent of the Milwaukee area's homes had washing machines and 66 percent had vacuum cleaners.

(One electrical "appliance" that TM did not sell was taking the country by storm: the radio. When the nation's first commercial station, KDKA in Pittsburgh, signed on in 1920, America

The PSB auditorium was regularly filled to overflowing for demonstrations of electric cooking.

jumped headlong into the age of mass entertainment. Milwaukee joined the craze in 1922, when WAAK began to broadcast from Gimbels Department Store. As home kits gave way to factory-built units, the proportion of local families who owned radios jumped from 9 percent in 1924 to 44 percent in 1927. Radios added to the power load, but power lines frequently interfered with radio reception. Complaints rose from 300 in 1924 to 1,288 in 1926, and a roving crew of TM troubleshooters worked full-time to eliminate sources of interference.)

Commercial accounts kept pace with growth on the residential side. By 1927 practically every business in the Milwaukee area was linked to the TM system, and any town that lacked a "White Way" in its business district was considered hopelessly backward. The number of advertising signs proliferated, with a preference for neon lights apparent late in the decade. Lighting also opened the door to night-time entertainment; in 1929 Marquette University played the city's first night football game under new lights at its stadium on the West Side. But the greatest growth in the commercial load followed a building boom in downtown Milwaukee. The Schroeder Hotel (now the Milwaukee Hilton) was one of the largest additions to the skyline. Dedicated in 1927, it consumed as much power as 6,000 houses.

Industrial users remained the system's bread and butter. Although some factories relied on independent power plants until well after World War II, TMER&L counted "most" of the city's industries among its customers in 1927. Some of the largest — Allis-Chalmers, Cutler-Hammer, Square D, Allen-Bradley — were in the electrical business themselves, producing everything from turbines to resistors for the national market. All industries (with the temporary exception of the breweries) were booming in the Twenties, and the result was a record surge in the demand for power. TM began to track the power needs of 30 large industries at mid-decade; their combined consumption rose 49 percent in 1927 alone and jumped another 69 percent in 1928. The case of Allen-Bradley, a specialist in industrial controls, is probably typical. When the company connected to TM lines in 1918, its monthly usage was a healthy 50,000 kilowatt-hours. With the completion of a major addition in 1928, Allen-Bradley's consumption soared to 200,000 kilowatt-hours a month, and it topped 258,000 only a year later.

The urban load — residential, commercial, and industrial — dwarfed all other sources of revenue, but S.B. Way did not ignore his system's smaller markets. Wisconsin Michigan Power and Wisconsin Gas & Electric expanded aggressively in their service areas, in part to keep the Insull and Byllesby forces at bay, but also to build their own loads. WG&E placed particular emphasis on gas sales, doubling its customer base in the 1920s. The company laid 250 miles of gas pipe in 1928 alone, reaching from Racine all the way to Watertown. TMER&L itself, whose borders with WG&E were ill-defined at best, sprawled well beyond the city. In 1925 the company

Wisconsin Gas & Electric stressed gas sales in the 1920s. WG&E crews laid hundreds of miles of new pipe, including this section on Water Street in downtown Racine.

WG&E billboards touted the advantages of "electric service" in rural Wisconsin.

launched a Rural Service Bureau whose stated purpose was "to extend the blessing of electric service to the thousands of wonderful farms in the territories [TM] serves." Five salesmen, including one who periodically made calls on skis, scoured the countryside in search of customers. In 1926, for instance, they drew 325 farmers and their families to a two-hour meeting in the Jackson (Washington County) town hall. The crowd saw ranges and toasters in operation, heard the EMBA band, watched all three reels of *Electrical Milwaukee,* and helped themselves to refreshments.

Farmers may have enjoyed the sales pitch, but they had to be convinced that electricity, like any other implement, would pay for itself. Well-publicized tests, including one showing that chickens laid 50 percent more eggs under artificial light, helped to convince the skeptics. The number of rural customers on the system's lines shot up from 280 in 1919 to 3,000 in 1925 and 8,222 in 1928. TM's annual report for 1928 noted a turning point: "Where previously the greater emphasis was placed on extension of new rural lines, it now is possible to devote sales activities primarily toward building the load on existing lines."

The system's growth in all markets and in all service areas was little short of astonishing. The power and light business had risen to new heights during James Mortimer's administration, reaching a level of 85,030 customers and 237.7 million kilowatt-hours in 1919. Those figures soon seemed like little more than a good beginning. Led by TMER&L, which accounted for more than 80 percent

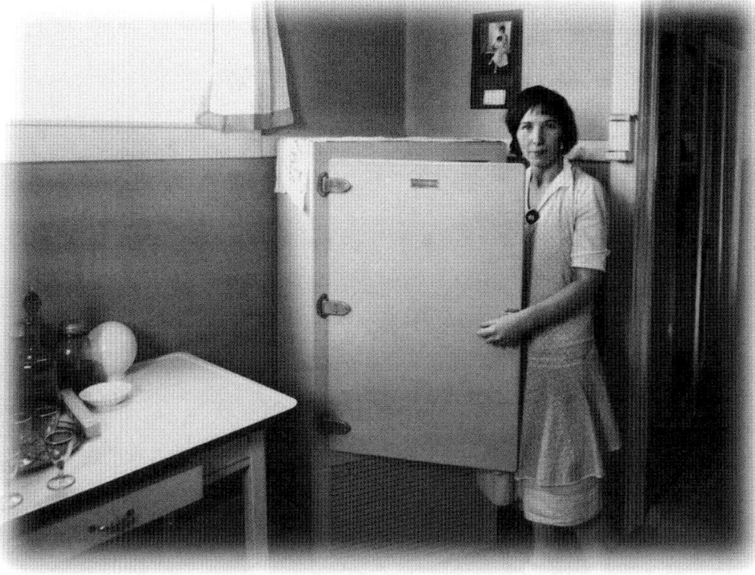

Farm families were quick to sign up. Jake Lasky of rural Seymour showed off the modern electric drive in his barn, while Mrs. Lasky posed with the new electric refrigerator in her kitchen.

of the total business, the system served 298,696 customers in 1930, and kilowatt-hours topped 1.27 billion. The combined utilities added nearly 20,000 customers every year during the 1920s, and energy consumption grew by a factor of five!

Customers and Constituents

There was a subtle shift in the TM system's public image during the Twenties. In his first term as president, John Beggs had not shown particular concern for the system's customers; he provided the best service he could, and the public, essentially, was welcome to take it or leave it. James Mortimer had worked to instill a strong customer orientation in his troops and, with the business growing at a record pace, S.B. Way further narrowed the gap between TM and its public. Way was the primary author of "home-town financing" in 1918, and he took other highly visible steps to make the system a customer-friendly enterprise. In 1926 he sponsored a system-wide open house that attracted 69,195 visitors. (Eighty-six percent toured the Public Service Building, and another 10 percent stopped at the Lakeside plant.) In 1928 Way opened Customers Hall on the second floor of the PSB. It was the last word in one-stop shopping: Customers could pay bills, order service, make complaints, apply for credit, and buy stock in one convenient location. TMER&L entered the Radio Age in 1929, launching the weekly *Kilowatt Hour* on WTMJ. Hosted by "Wattson Volts," the program featured "melodious selections of high character" played by the Kilowatt Concert Orchestra and guest soloists. At the beginning of every broadcast, Mr. Volts (Al Engelhard) intoned, "The Kilowatt Hour is striking, heralded by the chimes of the Telechron all-electric clock." Within a year, the show was drawing as many listeners as *The A&P Gypsies, The Firestone Hour, The Fleischman Sunshine Hour,* and other national "chain programs."

Radio shows, open houses, and first-rate electrical service helped to soften TM's old public image as a money-grubbing monopoly, but they did not erase it entirely. Politicians, in particular, continued to chafe at their inability to control the system. In 1921, for instance, Waukesha's city attorney began a highly publicized campaign to force TM to provide free soap and towels in the local interurban station's washrooms. (The campaign failed.) In 1923 Racine voters made utility opponent W.H. Armstrong their mayor; he promptly revoked a WG&E permit to lay new gas mains in

RAIL & WIRE
OCTOBER 1926
E.M.B.A. WISCONSIN
EMPLOYES' MUTUAL BENEFIT ASSOCIATION
Open House Week
Nov. 8th-13th
All Are Welcome
The Electric Co.
SERVICE FIRST

S.B. Way did his best to make TMER&L customers feel welcome. A 1926 open house attracted more than 69,000 visitors.

the downtown streets, and summarily arrested the construction crews. (Armstrong lost the ensuing court case.) Although other politicians got into the act, "Fighting Dan" Hoan remained the champion utility-basher in the region. Working with the socialist bloc on the Common Council, Milwaukee's mayor did all he could to make life difficult for TMER&L, platting vacant streets in the path of new transit lines, beating the drum for lower power rates and streetcar fares, and demanding that TM extend its single-fare zones into newly-annexed neighborhoods.

Shortly after his return to power in 1920, John Beggs had announced that a better relationship with Milwaukee's irascible mayor was among his highest priorities. Hoan clearly found the old man a worthy adversary, declaring him "concededly the shrewdest financier and dividend maker which the utility corporations have supplied in Wisconsin in recent years." The pair huddled in May, 1920, producing what the TM board considered "favorable indications of cooperation." Those expectations came to nothing; Hoan was soon back on the soapbox, berating TMER&L and all that it stood for.

The mayor had already struck a small blow for public ownership, which remained his long-term goal. In 1916 Milwaukee voters had approved a bond issue for a municipal street-lighting system, and the work consumed the next five years. New lights replaced a mismatched system of old lamps, including some genuine antiques; the city's last arc light finally sputtered out in 1926. The new system's signature was the celebrated "harp luminaire," an elegant fixture that has since become a collector's

The company opened Customers Hall, a full-service facility, in 1928 ...

... and hit the airwaves with The Kilowatt Hour *a year later. "Wattson Volts" handled the announcing chores.*

The City of Milwaukee operated a pint-sized powerhouse near the lakefront in the 1920s. It was the small brick building shown here between the stacks of the municipal garbage incinerator. (Milwaukee Public Library)

item. In 1921, as the last harp lights were going up, Hoan proposed a municipal power plant to run the system. (Milwaukee already had a tiny 600-kilowatt plant connected to its garbage incinerator. The plant's output was sold to TMER&L, making up precisely 0.28 percent of the system's capacity in 1927.) Anxious to retain the street-lighting load, TM officials offered the City a rate so low that Hoan dropped his power plant proposal immediately. The mayor, however, was soon crowing that his threats had saved voters "over $100,000 per year in the price of purchased energy," and he demanded that the company make comparable reductions in its commercial rates.

Larger issues of ownership came to a head in a 1925 referendum. Six years earlier, the Common Council had appointed a Municipal Acquisition Committee to study the feasibility of taking over TM's streetcar and electric properties. Headed by Fred Hunt, president of the Milwaukee Dustless Brush Company, the multi-partisan committee studied the issue for years, concluding that an outright purchase was "purely an academic question" under the prevailing legal and financial circumstances. The group proposed a complex alternative hammered out in negotiations with TM officials. The heart of the plan was a "service at cost" contract that would have specified service levels, power rates, and streetcar fares agreeable to both parties. (The contract was intended to supersede Railroad Commission control, and it included, significantly, a guaranteed 7.5-percent return on capital.) Although North American retained ownership under the plan, the City had the option of investing in the system, with all investments applied toward the eventual purchase of TM's Milwaukee-area properties for $50.7 million.

The major newspapers lined up behind the proposal. "Of course we shall pay for service," wrote the *Milwaukee Journal,* "but here is a contract which allows the city to say what service it will elect to have and how far it is willing to go in paying for it." From Dan Hoan's point of view, the plan gave Milwaukee a greater voice in the system's operation and the possibility, however remote, of municipal ownership. (The price tag was nearly twice the City's entire $28 million budget in 1925.) From TM's point of view, the proposal offered a 7.5-percent return and the promise of what S.B. Way called "better cooperation" with local officials.

From the voters' point of view, the whole thing was a bad idea. On April 7, 1925, Milwaukeeans rejected the plan by a two-to-one margin. Although the proposal's aims were considerably more modest, the vote was widely viewed as a referendum on municipal ownership. Their par-

Daniel in the lion's den: Mayor Dan Hoan conferred with S.B. Way at the company's 1926 open house.

ents may have gleefully boycotted the streetcars during the 1896 strike, but the city's voters had apparently grown content with the status quo. Some supporters argued that the referendum question was poorly worded and subject to multiple interpretations. "The people were misled," declared Dan Hoan and Fred Hunt. "Misrepresentation ran wild," concluded the *Milwaukee Journal*. TM officials, who may have secretly counted on the referendum's defeat, considered the result a vindication. *Rail & Wire* called the outcome "a vote of confidence" that affirmed the system's ability to function "without additional political supervision." Noting that voters in both Milwaukee and Chicago had turned down chances to control their systems, the *Electric Railway Journal* (April 11, 1925) applauded the trend: "The municipal ownership hobby in these cities at least is apparently joining the remainder of the equine family as an obsolete factor in municipal transportation."

Although Dan Hoan accepted the verdict, his public animosity toward TMER&L continued unabated. His stands were undoubtedly based on deeply held principles, but there was also a certain amount of posturing in Hoan's anti-utility crusade. Like many career politicians, he could be as conciliatory in private as he was confrontational in public. The mayor appointed TM executives to municipal boards, attended events like the 1926 open house, and enjoyed generally cordial relationships with Beggs and then Way. From a political viewpoint, however, TMER&L was a convenient enemy, a monopoly with enough accumulated baggage to serve as the ideal foil for a fiery populist. It was his battle with TM, as much as any other crusade, that defined the mayor's public image as "Fighting Dan."

Rubber Tires and Rapid Transit

If political controversy was one problem that refused to go away in the 1920s, the transit system's general anemia was another. Patterns that had surfaced in the Mortimer years intensified: stagnant ridership, lagging revenue, and a lack of confidence in the system's future. TM operated one of the largest and most comprehensive rail networks in the nation, but passenger traffic showed practically no growth in the 1920s, beginning the decade at 212 million riders and ending it at 216 million. Gross income was just as disappointing, hovering around the $10 million mark in most years. TM's firstborn child became a problem child, particularly when its performance was compared with the bright record of the power and light business. The electric side had surpassed the transit operation as the system's profit center in 1914, and its lead grew steadily. TM's railways averaged a return of 4 to 5 percent during the Twenties, while the electric business typically generated twice that figure.

In their public pronouncements, TM officials were careful to underline their commitment to the transit system. In 1922, when a Milwaukee alderman called the city's street railways "a lot of junk," TM placed an ad declaring that "next to the city water system the railway is the most vital of all public services to the life of the community." In 1927 *Rail & Wire* published photographs of a new line on Atkinson Avenue as "proof positive that development and expansion keep time to the tune of the motorman's gong, and that the wheels of progress are under the street car."

In closed session, however, con-

Despite stagnant ridership, TMER&L trolleys carried millions of passengers annually.

The fareboxes they filled were duly emptied and the results tallied by accountants in the Public Service Building.

Track renewal projects kept this asphalt crew busy in 1921. "We take considerable pride," wrote Rail & Wire, *"in the quantity and quality of the work that has been done this year by this gang."*

cerns were freely expressed. At a 1921 meeting, the board noted that "the railway utility is not now and has not for years past achieved a reasonable return." In 1926 TM's Executive Committee groused that the transit operation was both more expensive and less profitable than the power and light business: "The track reconstruction program is requiring new investment approximating $1 million per year, and the proportion of the return on railway property being carried by the electric service is constantly increasing."

One of the few bright spots in the transit picture appeared in 1927, when Schuster's Department Store inaugurated its annual Christmas parade on TM's rail lines. The parade became a cherished Milwaukee tradition, with a memorable, and fondly remembered, cast of characters. Billie the Brownie, Santa's "faithful helper and advance agent," described his boss's progress from the North Pole in daily radio reports. The weeks of waiting always reached a climax on the Saturday after Thanksgiving, when a procession of lavishly decorated TM flatcars left the Cold Spring shops for a seven-mile journey through the city's streets. The cars featured characters like Cinderella and Peter Rabbit, but the high point was the float carrying Santa Claus and six live reindeer, who were tended by "a real Alaskan Eskimo" named Me-Tik. (Although Me-Tik remained, stuffed specimens eventually replaced the live reindeer.) More than 300,000 children of all ages bundled up for the Schuster's parade every year, and it was, for an entire generation of Milwaukeeans, practically synonymous with Christmas.

Billie the Brownie and Me-Tik the Eskimo could neither solve nor hide the transit system's underlying problems. Although the constant sniping of politicians like Dan Hoan and the expensive mandates of the Railroad Commission aggravated the system's distress, the real problem was the riding public. America became an automobile-saturated society in the 1920s. The number of cars produced in the nation's auto plants soared from 181,000 in 1910 to 1,905,560 in 1920 and 4,587,400 in 1929. A surprising number were Fords. From its debut in 1908 to its demise in 1927, Henry Ford's Model T was easily the most popular car in America; in 1923 it constituted 57 percent of the country's total car production. Sold for as little as $260 (roughly $2,100 in 1996 dollars), the "Tin Lizzie" was fabled for its

Christmas began to ride the rails in 1927, when Schuster's held its first parade on TM's trolley lines. Santa, Me-Tik, and their reindeer were photographed at Hanover Street School soon after the parade's debut.
(Milwaukee County Historical Society)

simplicity and operating economy. In 1922 TMER&L employees who used their Fords on company business were reimbursed 3.9 cents per mile; owners of all other cars were paid 5.4 cents.

As the Model T and its counterparts took the country by storm, private transportation eclipsed public transit. In Wisconsin, growth in the number of cars per family was dramatic: one in 84 in 1910, one in two in 1920, and one for every family in 1930. Wealthier households, including S.B. Way's, undoubtedly had more than one vehicle at their disposal, but automobile ownership had become nearly universal. As the "good roads" movement gathered momentum, there was corresponding growth in the state's highway system; Wisconsin's trunk highways covered 249 miles in 1920

The growth of automobile traffic posed a grave threat to TM's transit system, and competition for the right-of-way was most intense in downtown Milwaukee. The view is east on Wisconsin Avenue from Third Street in 1926.

and 2,606 miles only a decade later. Affordable cars and up-to-date roads created an alternative transportation system, one whose advantages over public transit — convenience, flexibility, independence — were self-evident. The streetcar retained its hold only in the larger cities, and then largely among commuters.

TM transit officials had no difficulty naming their enemy. In 1921 John Beggs ascribed the drop in streetcar revenue to "automobile competition." In 1923 *Rail & Wire* lamented the dearth of warm-weather travelers: "There was a time and that but a few years ago when the street railways were hard pushed to handle their summer crowds, but Henry Ford has made those days a memory." In 1926 TMER&L's directors noted that ridership was rising in rush hours and falling at other times, "a natural consequence," they concluded, "of the use of private automobiles." A TM traffic study showed that there were 18 automobiles for every streetcar on Wisconsin Avenue in 1926, even though streetcars carried four times as many passengers. Writing in the *Electric Railway Journal* (Feb. 6, 1926), TM transit manager Roy H. Pinkley observed, quite properly, that the competition was unfairly subsidized: "The same legislatures that have taxed the interurban railways as monopolies have constructed highways with public funds and permitted the unrestricted operation of competitive bus and truck service over such roadways."

Comparable public support for a private utility was not on anyone's agenda. TM's sole alternatives were to preserve the traffic it had and to cut costs where it could. Labor was the system's major expense, and two labor-saving innovations appeared in the Twenties. The first was the two-car, three-truck "articulated" train, built in the Cold Spring shops from old streetcars. Its designer was S.B. Way himself, who maintained a strong interest in the technical side of the business. Used exclusively in rush hours, Way's "Siamese trains" debuted in 1920, and by 1925 there were 125 units in service. The second innovation was the one-man "safety car," which appeared on the 27th and 35th Street lines in 1921. (Conductors on the remaining lines were immediately downgraded to "collectors.") Despite the spirited objections of public officials and

Trying diligently to trim its transit expenses, TMER&L introduced the labor-saving "articulated" train, an S.B. Way invention, in 1920.

The one-man, front-entrance "safety car" followed in 1921. (Russell Schultz)

The 900-series trolleys, like this one-man car shown on a training run, were the last additions to the Milwaukee fleet. The final shipment arrived in 1929. (Russell Schultz)

local residents, conversion of all cars to one-man operation was well under way by 1930.

Cost-cutting also applied to rail extensions; TM made it abundantly clear that the system was not in a growth mode. In 1921 the Executive Committee adopted a policy of "resisting to the utmost applications for rail extensions unless such extensions are financed for the Company." When Prohibition prompted the Uihlein family, owners of Schlitz Brewing, to build a candy-bar plant north of the city in 1922, they paid TM $20,000 for the privilege of streetcar service. Resistance to growth was accompanied by selective retrenchment. Hoping to reach Janesville and Madison, John Beggs had laid short sections of track beyond the original interurban terminals in Burlington and Watertown. TM decided to tear out those stubs in 1923, an open admission that the interurban system would grow no more. The same policy applied in Milwaukee; streetcar tracks on Oklahoma Avenue were removed when a planned extension was canceled in 1925.

As in earlier years, TM's experience mirrored national trends. Annual ridership on America's streetcars was absolutely stagnant between 1917 and 1929, despite a surge in population, and rides per capita plummeted from 260 to 226. During the same period, the nation's electric railway trackage dropped from 45,000 miles to 38,000, and the number of passenger cars in service fell from 81,000 to 73,768.

Economizing in some areas and modernizing in others, TMER&L tried to beat its competitors in the new world of motorized transport, but the company also attempted to join them. Truck service from interurban railheads expanded steadily into the hinterland, reaching Lake Geneva, Delavan, Whitewater, Madison, Beaver Dam, and Fort Atkinson in 1922. Operated as rail feeders at first, the truck lines helped to bring in one-fourth of Milwaukee's fresh milk supply and large quantities of poultry, eggs, and veal. A novel experiment took TM even closer to the competition. In 1928 Badger Auto Service, a corporate shell established five years earlier, began to operate "commercial automobile service stations" in the Milwaukee area. Its largest property was the old Pabst light plant (and former TM headquarters) on Broadway, which became a well-used parking structure, filling station, and tire store for downtown commuters. Badger's other operations included a parking lot on Second and Michigan (still owned by Wisconsin Electric and still a parking lot) and two full-service filling stations on 35th Street, one at Fond du Lac Avenue and the other at

There were also accommodations to the motor age. Company trucks transported much of the region's fresh milk ...

... and TM's Badger Auto subsidiary operated filling stations like this one at 35th and Wells Streets.

(top photo: Russell Schultz)

Wells Street. Although TM's directors soon concluded that their foray into the automotive world "does not seem to be good policy," customer demand kept Badger Auto's establishments in operation for years.

The most important sign of accommodation to motorized transit, and by far the most enduring, was the rapid expansion of TM's bus service. In the years after World War I, buses were every bit as novel as electrified streetcars had been 30 years earlier, and their evolution was just as rapid. The first models were little more than oversized automobiles, with an average capacity of perhaps seven passengers. As frames, suspensions, tires, and motors improved, the average capacity rose to 20 in 1920 and 40 in 1927. By

Buses made up some of the ground that streetcars were losing. This 38-passenger Twin Coach model began to operate in 1927.

1929 "streetcar-type buses" had made their debut, with enough room for 50 riders. "Any equipment more than one year old," wrote TM's Roy Pinkley in 1926, "might be considered out of date." Despite their rapid obsolescence, buses were considered the wave of the future in some circles. In 1920 one manufacturer published an ad announcing, somewhat prematurely, that the bus had "sounded the death knell of the street car," which was dismissed as an "awkward" conveyance forever tied to "unsightly and abominable" wires. "The motor bus," the ad proclaimed, "is the poor man's limousine. It gives him all the comforts and delights of closed-car motoring with none of its expense. It is swift, sure, silent, comfortable, and even luxurious."

TM was among the first transit systems in the nation to add motor buses to its fleet, and they became a major source of revenue on both city and interurban lines. The company did not, however, enter the field voluntarily. In nearly every case, bus service was launched as a direct response to competitive threats, which surfaced with nagging regularity. In 1920, when a Minneapolis firm began to run buses parallel to TM's Milwaukee-Racine-Kenosha interurban line, the company decided to fight fire with fire. "The situation was squarely put," recalled S.B. Way in 1925, "as to whether the electric railway should lie down and take a good licking, or whether it should stand up and make an effort to remain the predominant factor in organized transportation service." Way was not about to take a licking, and the company adopted an uncompromising, even ruthless, defensive strategy. Wherever competitors emerged, TM ran buses on the same routes immediately, adding trips and undercutting fares until the enemy was forced to throw in the towel. The approach was generally successful. In 1922 TMER&L bought out two ruined competitors for $84,000 — roughly the value of their rolling stock and terminal facilities. TM maintained its dominance, Way bragged, "without paying tribute."

Once competitors had drawn TM into the fray, the company decided that rubber-tired vehicles might have a permanent, if secondary, place in its transit system. Milwaukee's city service began with a short feeder on Mitchell Street in April, 1920. By 1929 "poor man's limousines" navigated a total of 18 routes covering 53 miles in the city; all offered free transfers to streetcars. Some lines were openly targeted to affluent commuters who had seldom seen the inside of a trolley. In 1923 TM inaugurated its "Green Bus" service in the money-colored neighborhoods on the East and West Sides. Passengers paid a dime to ride the bus downtown (three cents more than streetcar fare), and the rolling stock included several highly visible double-deckers. The sheer novelty of the new buses, which remained in use until 1932, attracted a steady stream of pleasure riders. "The upper deck of a motor bus," reported *Rail & Wire* in 1923, "has over night become the favorite vacation playground of Milwaukee's younger set. It is both romantic and inexpensive."

Interurban bus service grew even more rapidly than the city lines. After its first modest experiments in 1919, TM added routes at a dizzying pace, generally extending the rail network John Beggs had built at the turn of the century. By 1921 the company was running five buses to Racine, Waukesha, and other points where competition had blossomed. In 1922, when Wisconsin Motor Bus Lines took form as an operating division of TMER&L, there were 54 interurban buses, and they traveled as far as Madison, Fond du Lac, Janesville, and Beloit. The Plankinton Arcade, just up the block from the Public Service Building, became the Wisconsin Motor Bus system's terminal. It was also the home of perhaps the first tour bus service in the state; between 1924 and 1926, the "white and gold fleet" carried sightseers to all the scenic wonders of Wisconsin, from Devil's Lake to Door County.

Despite the steady growth of rubber-tired transit, both urban and interurban, TMER&L did not become The Milwaukee Bus and Light Company. Electric cars continued to carry the

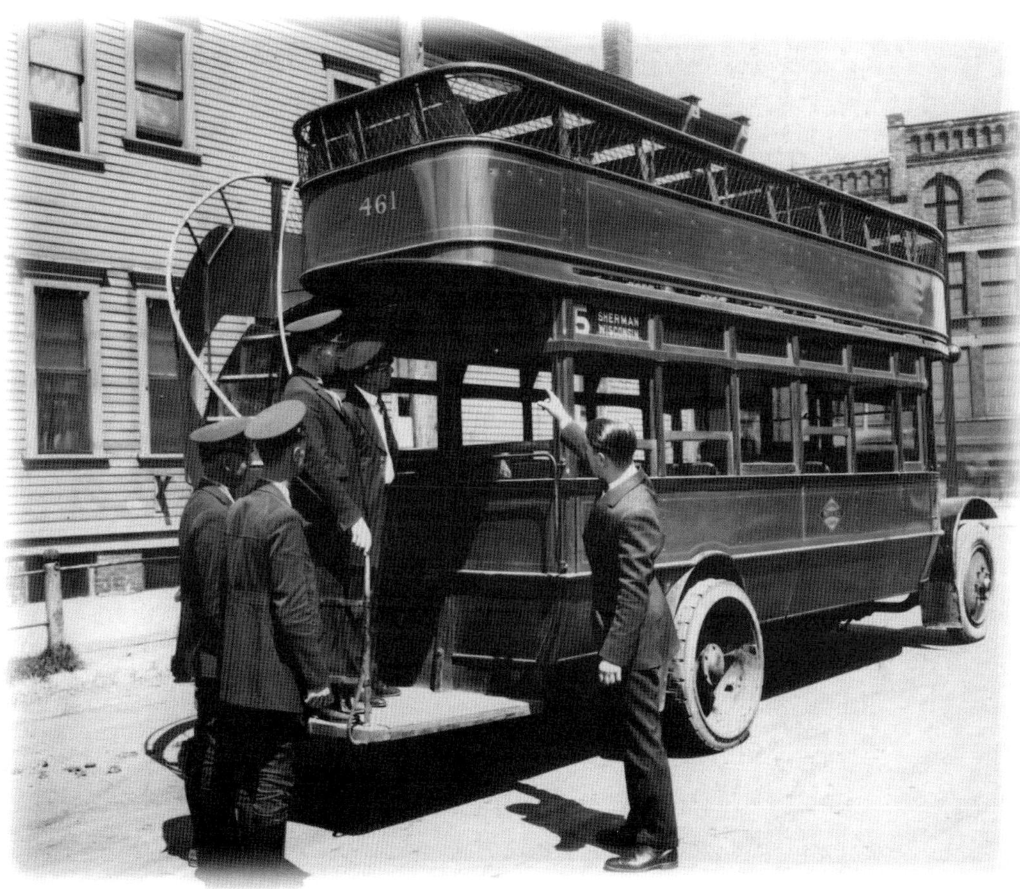

Double-deckers added a welcome touch of novelty to a relatively static transit system ...

... and "white and gold" buses like this 1924 Pierce-Arrow carried tourists to scenic attractions throughout Wisconsin. (Milwaukee County Transit System)

Interurban bus service showed the most dramatic growth. In 1923 all of TM's interurban buses lined up for a group portrait in Milwaukee's Juneau Park.

vast majority of the system's riders. In 1927 TM operated 916 passenger cars that traveled a total of 65,000 miles every day; in the same year, 150 buses put on 9,000 miles a day. Company officials insisted that buses were only low-cost supplements to the rail network, serving as feeders, extensions, and safeguards against competition. In 1920, for instance, James Mortimer pronounced buses ideal for "sparsely settled and suburban districts," where they could develop traffic "to the point where the large investment in a rail line may be justified." In 1921 the Executive Committee declared that bus service was TM's "only weapon to protect its heavy investment in railway." A 1922 company ad expressed boundless confidence in steel-wheeled transit: "Motor buses will never supplant electric railway cars in large cities. They will be increasingly useful as feeders to electric car lines and as supplementary lines."

The Watertown terminal became a point of transfer between trains connecting with Milwaukee and buses traveling to or from Madison, Fort Atkinson, and Beaver Dam.

Although the railways remained paramount, buses had some incontestable advantages. They offered a freedom and a flexibility streetcars could never match, and they required a fraction of the capital investment. Establishing rail service was like building a house, while initiating bus service was as easy as pitching a tent. Rubber-tired transit was also, until the mid-1920s, completely unregulated.

Buses, therefore, soon became an end in themselves. TM's bus lines were a million-dollar operation by 1925, and numerous routes had no connection whatsoever with existing rail lines. Buses were the vehicle of choice on the

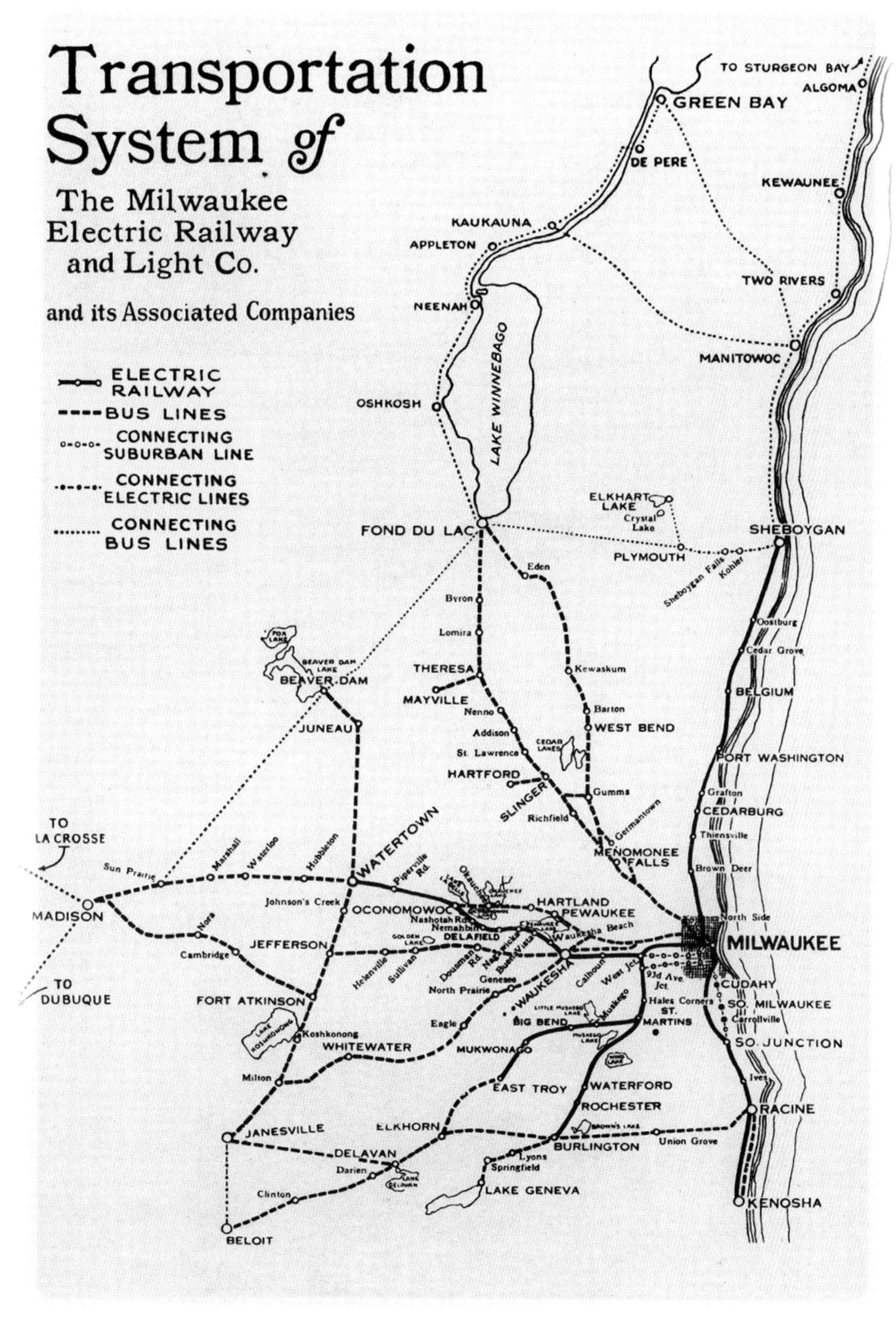

By bus and by rail, TMER&L nearly covered southeastern Wisconsin with transit lines in 1929.

developing margins of Milwaukee, and interurban service reached as far as Green Bay, Sturgeon Bay, and (in 1930) Iron Mountain. The bus business was profitable from the start, and rubber tires rapidly gained ground that steel wheels were losing.

The interurban railways were losing ground with the greatest speed; their share of the system's traffic had fallen to an anemic 2 percent by 1921. In the next year, despite steep fare cuts intended to boost ridership, the interurban cars lost $130,244, while interurban buses grossed $228,814. The problem was especially acute in out-state Wisconsin. In 1928, after years of substantial losses, Wisconsin Michigan Power abandoned its interurban line between Neenah and Kaukauna. Appleton's streetcar service was the next to go; one of the oldest traction systems in the country became a memory on April 6, 1930. (Passengers dismantled the last car after its final run, carrying away money boxes, sand pails, and light bulbs.) Bus service immediately replaced railways in the Fox Valley. Although it was only the beginning of the end, TM's rail empire was showing definite signs of contraction.

The local experience was disappointing, and the national trends were dismal: Ridership on America's interurbans dropped 50 percent in the 1920s. Despite abundant evidence that it was a risky idea at best, TMER&L decided to attempt a multi-million-dollar resurrection of its interurban network. It is likely that John I. Beggs, the erstwhile transit king, was behind the decision. Although he made no attempt to reverse the trend away from traction in TM's general plans, the interurbans were a cherished

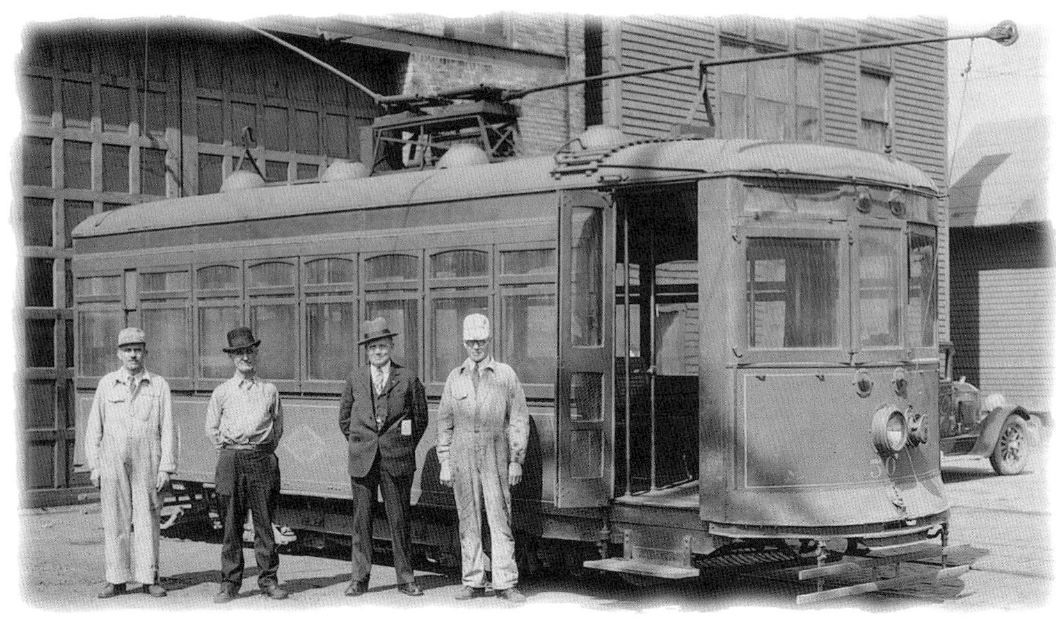

Company veterans posed with Appleton's last streetcar on the morning of its final run in 1930. Souvenir-hungry patrons practically dismantled the car by day's end.

exception. At the beginning of his second presidency in 1920, Beggs had expressed a desire "to further develop the interurban lines and to make them attractive properties." In 1921, as a first step, TM inaugurated limited express service on its Watertown line; the company hoped to match the speed of steam railroads at much lower fares. In 1924 the Cold Spring shops began to rebuild old interurbans with an emphasis on elegance; some of the refitted cars featured carpeted floors, smoking compartments, observation platforms, and seats of Spanish leather.

Beggs' plans reached their peak in 1925, when TMER&L began work on a "Rapid Transit" corridor linking Milwaukee with all points west. The heart of the project was a new right-of-way covering five miles between 35th and St. Paul on the east and 100th and Burnham (West Junction) on the west, where it met the Waukesha-Watertown interurban tracks. Threading its way between West Allis and Wauwatosa, the new corridor was basically a short cut; it enabled interurban cars to avoid miles of congested city and suburban streets. At $1.6 million (including a new bridge over the Menomonee River), the project was expensive, but it signified a renewed commitment to the interurban system that had long been Beggs' pride and joy.

John Beggs died months before the work was finished. Somewhat surprisingly, given the moribund state of the interurban business, S.B. Way pushed the line through to completion. "The project is distinctly building for the future,"

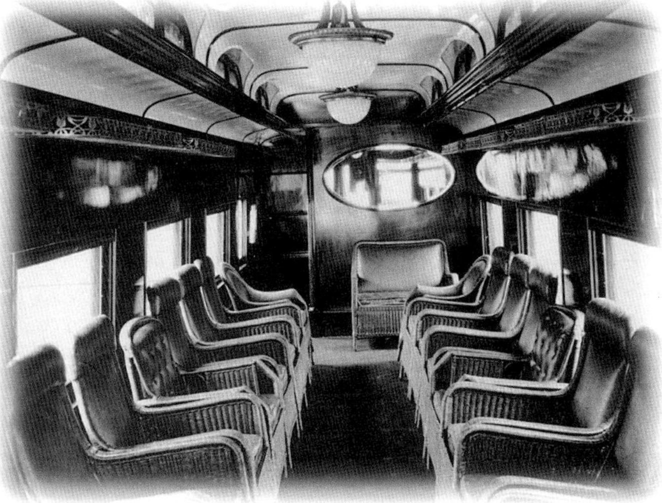

Bucking some rather ominous trends, John Beggs tried to lure back interurban passengers with express service and elegantly appointed cars like the Watertown Limited on the left. Some of the refurbished parlor cars (right) featured mahogany trim, soft leather seats, and lights that provided "abundant yet soft illumination for comfortable reading." (left: Russell Schultz, right: Milwaukee County Transit System)

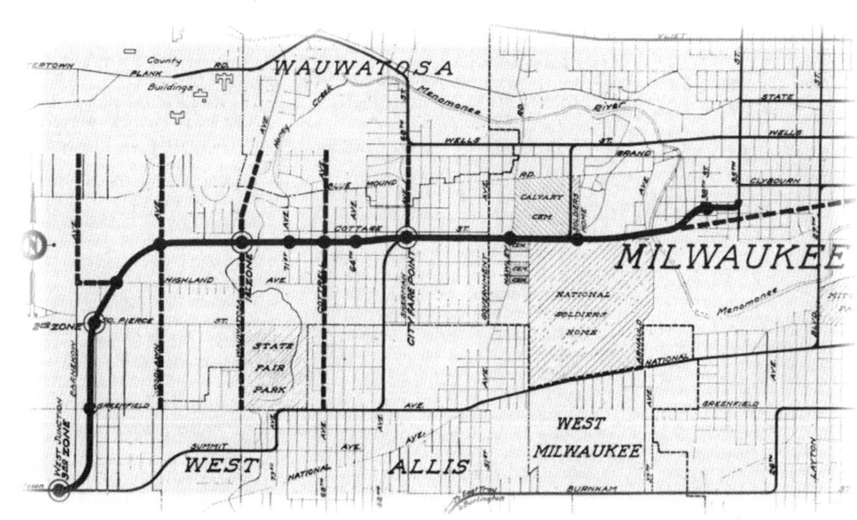

The Rapid Transit project, a high-speed shortcut between Milwaukee and all points west, promised "the most modern electrical railway service in the country today."

It required painstaking survey work ...

... a deep cut in the vicinity of Greenfield Avenue ...

... and a new viaduct over the Menomonee Valley near the Pigsville neighborhood. Interstate 94 follows the original Rapid Transit route today.

Way said in 1926, "and evidences the faith of the Company in the continued growth and prosperity of Milwaukee and its environs." The corridor opened for service on June 14, 1926. With larger, faster cars and five miles of new express track, the Rapid Transit line provided 35-minute service to Waukesha — close to automobile time. Ridership jumped 40 percent immediately, and traffic to Watertown was up 66 percent by year's end. The *Electric Railway Journal* (November 20, 1926) found the results heartening: "High speed and frequent service, coupled with the use of the latest car equipment, make it possible for an interurban successfully to cope with the private auto and bus."

The experiment was successful enough, in S.B. Way's judgment, to merit further investment. In 1927, with the completion of a new cut-off at West Junction, all Burlington and East Troy cars began to operate on the Rapid Transit corridor; their running times were reduced by 23 minutes. In 1928 a comparable short cut around South Milwaukee enabled Racine cars to reach their destination 19 minutes faster. A TM-built dining car was added to the Watertown line in the same year, bringing interurban service up to the best standards of the steam trains. Bridge crossings on the "Milwaukee Electric Lines" were soon adorned with signs boosting "Rapid Transit: Speed, Comfort."

In 1928 TMER&L added a dining car to its Watertown interurban train. Passengers could order everything from smoked tongue to stewed tomatoes. (Milwaukee County Transit System)

One major bottleneck remained: downtown Milwaukee. Interurban cars might have traveled at top speed in the countryside, but they slowed to a crawl in the heart of the city. As the terminal for the entire system, the Public Service Building was a focal point of congestion; passenger trains, express freights, TM buses, and private automobiles competed for the same space, creating a condition approaching gridlock. In January, 1928, TMER&L announced that relief was in sight. System officials unveiled plans for a 2.8-mile corridor linking the PSB with the existing Rapid Transit corridor at 40th Street. Looking ahead to the suburbanization of Waukesha County, S.B. Way called the extension "the essential downtown link for east and west traffic between the heart of the city and the great area to be developed to the west." In an unprecedented burst of public approval for TM, Mayor Dan Hoan declared the venture "a wonderful thing" and "a valuable birthday present" for the citizens of Milwaukee.

The downtown link was budgeted at $4 million, more than twice the cost of the first Rapid Transit corridor for half the mileage, but the project had some unusual features. TM's engineers perched most of the corridor halfway up the Menomonee Valley bluff, an approach that required steep retaining walls in some sections and elevated tracks in others. Workers ultimately removed 525,000 cubic yards of dirt and stone, poured 19,000 yards of concrete, and demolished 150 buildings. The project also called for two imposing new structures. The first was an eight-story freight terminal at 10th and St. Paul, designed to relieve traffic congestion around the PSB. The second, and certainly the more unusual, was a five-block-long "subway" between the PSB and Eighth Street.

Construction began in December, 1928, and most of the work was completed. Soon enough, however, the Depression brought the project to an abrupt halt; Milwaukee's only subway was stillborn. Rapid Transit was, in hindsight, the interurban's last hurrah, and it ended with barely a whimper. Thirty years later, the corridor would be appropriated for a new version of rapid transit: Interstate Highway 94.

Company crews worked on a subway — Milwaukee's first and only — that would have connected the Public Service Building with the Rapid Transit tracks at Eighth Street. The project was a casualty of the Depression.

The 1920s were the last of the formative decades for the TMER&L system. In a time of swift change and feverish expansion, the system went through a series of transitions: from steel wheels to rubber tires, from political turmoil to a tentative truce and, most importantly, from local service to regional superpower. In every case, the transitions took TM to the threshold of the present. Major developments of the Twenties, from the rise of pulverized-coal power plants to the advent of a national power grid, are still very much with us. The system's service area had assumed its present shape by 1925 and, with the death of John I. Beggs, the age of self-styled entrepreneurs gave way to the era of professional managers. Despite bold holding actions like Rapid Transit, TMER&L's identity as "the Electric Company" crystallized in the 1920s. The balance of power swung decisively from traction to electric service, and the system took on an image and a reality that have changed little, in their essentials, to the present. There have been manifold changes since the Twenties — the streetcars are gone, and the North American Company is a memory — but Sylvester B. Way would most certainly recognize the Wisconsin Electric system of today.

The 1920s were also, quite arguably, the last of the formative decades for contemporary American culture. The United States emerged as an identifiably modern society by 1930; its citizens were entertained by mass media, transported by private automobiles, bombarded by consumer products, and oriented to the city. New technology powered much of the transformation, and electricity powered key elements of the new technology: radios and refrigerators, movies and microphones, assembly lines and other agents of pervasive change. Electricity, in short, became an American institution in the Twenties; the Jazz Age was also a watershed period in the Electric Age. The utility business was no longer a revolutionary experiment but a central force in the life of the nation, and it would thereafter contract and expand with American society as a whole.

Chapter 6

1929-1945

Harsh Interlude

The stock market crash of October, 1929, ushered in the most extraordinary period of contraction and expansion in American history. The nation's economy first came apart at the seams, exposing an entire generation to unprecedented hardship, and then, with the outbreak of war, came together again in an unprecedented display of productive might. The twin cataclysms of the Depression and World War II dominated a dizzying, distinctly abnormal period spanning fifteen years. The reckless expansionism of the Roaring Twenties became a fading memory, replaced by questions about American capitalism that were answered, finally and emphatically, in mobilization for global conflict.

Although its highs and lows were less extreme than those that convulsed other industries, The Milwaukee Electric Railway and Light Company experienced all the tumult of the period. S.B. Way and his associates were forced to take a reactive stance, struggling to avoid red ink in the Thirties and to keep up with soaring demand in the Forties. Day-to-day management was more than enough to keep the executives busy, but the company also changed significantly as the result of two momentous policy decisions. The first was a conscious choice to leave the transportation business. The second decision was made by outside authorities. Political turmoil, a local constant for decades, reached the national arena, resulting in the untimely demise of North American

The Public Service Building, shown here on a May afternoon in 1936, served as TMER&L's anchor during a tumultuous period of economic struggle and military strife.

and its fellow holding companies. TMER&L left the period under both a new name — the Wisconsin Electric Power Company — and new, substantially local, ownership.

Hard Times

TMER&L closed the Twenties with a roar. In 1929 the company reported sales of 1.17 billion kilowatt-hours and net income of more than $5.3 million. Both figures were records that stood for years; profit levels, in particular, would not see comparable heights until after World War II. A downturn was obvious by 1930, when TM began to ration work and rotate crews in an effort to maintain full employment. In 1931 the company's output fell below the billion-kilowatt-hour mark for the first time in several years, and its slide continued to a low of 854.4 million kilowatt-hours in 1932 — 27 percent below the record of 1929. Earnings dropped even more precipitously; TM's net income tumbled from $5.3 million in 1929 to $1.3 million in 1934, a decline of 75 percent in just five years.

After decades of spectacular growth, the decline was shocking, but it clearly reflected local conditions. In 1932, the hardest of the hard years, Milwaukee's unemployment rate exceeded 40 percent, and half the city's property taxes went unpaid. Idle factories used no power, and idle workers used no transportation. As the dominant purveyor of both power and transit service in the region, TMER&L felt the impacts directly. Demand fell sharply, and consumers found it difficult to pay for even reduced levels of service; by January, 1933, the company was carrying 47,178 delinquent accounts, representing more than 20 percent of its customer base.

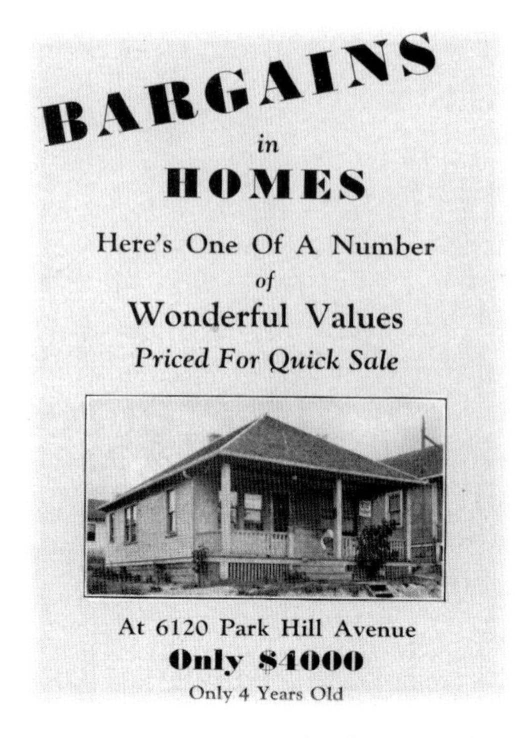

Every Depression-era issue of Rail & Wire *advertised homes "priced for quick sale" by the Employes' Mutual Saving, Building & Loan Association.*

The deepening gloom caused an abundance of casualties within the company. A disheartening round of layoffs reduced TM's employment from 6,538 workers in October, 1929, to 4,634 in June, 1933 — a drop of 29 percent. (Laid-off employees continued to receive EMBA medical benefits.) Workers without steady paychecks could not afford to keep up their mortgage payments, and the Employes' Mutual Saving, Building & Loan Association became the unwilling owner of hundreds of homes. The EMSB&LA advertised "unusual values in flats, cottages or bungalows ... priced for quick sale," including $14,000 homes offered for $10,000. Workers who managed to hang on to their jobs saw wages cut by 10 percent in 1932; even S.B. Way's salary was reduced, from $35,000 to $31,500. Dividends on common stock, all of it owned by the North American Company, were suspended in the same year. The all-company picnic, an annual affair since 1906, was held for the last time in 1931. *Rail & Wire* shrank to a four-page periodical in 1932 and appeared only twice in the next two years.

TM's sister companies shared the pain. Wisconsin Gas & Electric's revenues dropped alarmingly, and the industrial market for its gashouse coke disappeared. WG&E's only coke customer was the City of Racine, which bought tons of the fuel for distribution to needy families. Wisconsin Michigan Power was hit even harder. With a two-year supply of ore on the docks at Escanaba, dozens of mines in Upper Michigan's iron country were shut down, some permanently. By mid-1933, 35 of the 39 communities along the Wisconsin-Michigan border were unable to pay their street-lighting bills.

Although the TM system was, like most of its customers, under severe stress in the 1930s, the utility was never on the verge of collapse. As a tide of red ink swamped other industries, TM remained a solvent, even profitable, enterprise; the fact that it *had* a net income made the enterprise relatively unusual. The TM system's comparative health was the result of two factors. The first was its status as a regulated utility. State authorities set its rates, mandated its service standards, and allowed it a reasonable return, all of which provided an umbrella that no truly private industry enjoyed. TM officials may have griped about Railroad Commission decisions in earlier years, but state regulation was a godsend in the Depression.

The system's second inherent advantage was the nature of its product. Electricity had become a public necessity in the previous two decades; American society was wired to the power lines of the nation's utilities. Government relief efforts underscored just how important electricity had become. Nearly 20 percent of Milwaukee County's residents were on relief by 1933, and local agencies provided them with the bare essentials: food, clothing, rent, and utilities, including electricity. One monthly check from Milwaukee County to TMER&L kept thousands of families supplied with light and power. In the mid-1930s, when New Dealers built suburban Greendale as a garden community for central-city residents, they provided each rental unit with an electric range, an electric refrigerator and, the latest wrinkle, circuit breakers.

Electric service was a standard feature in every new home built in Greendale, a low-income "greenbelt" community southwest of Milwaukee. (Milwaukee Public Library)

Insulated by regulatory bodies and supported by the public's need for its product, the TM system did not experience the same level of hardship that practically flattened other sectors of the economy, particularly manufacturing. TMER&L's layoffs and wage cuts were relatively modest. The utility continued to pay dividends on its preferred stock, which was held by more than 23,000 individuals (90 percent of them Wisconsinites). Although austerity was their watchword, the TM companies were strong enough to take advantage of the depressed market for electric properties. In 1932, for instance, WG&E acquired the West Bend utility, adding 3,200 customers to its lines, and in 1934 Wisconsin Michigan purchased the Big Quinnesec Falls hydro plant from a mining company that was teetering on the edge of bankruptcy.

The TM system picked up a few bargains during the Depression, including the Big Quinnesec Falls hydro plant on the Wisconsin-Michigan border.

Hats off to progress: TMER&L officials and local leaders broke ground for the Port Washington Power Plant on May 26, 1930. Mayor August Kruke was at the controls.

The clearest sign of the system's relative health was the new Port Washington Power Plant. In 1929, certain that it would need more generating capacity to meet growing demand, TMER&L purchased a large tract of harbor land in Port Washington, a picturesque village 25 miles up the Lake Michigan shoreline from Milwaukee. The site offered an unlimited supply of condensing water as well as an anchorage for coal ships. (The Lakeside plant was supplied by train.) Ground was broken on May 26, 1930, months after the stock market crash, and the plant was scheduled for completion in 1932. The worst years of the Depression intervened, but construction crews stayed on the job. Their work involved leveling a 110-foot hill and using it as fill for a mammoth coal dock and storage area. Although there was no urgent need for its capacity, the first of five 80,000-kilowatt units was completed in 1935 and dedicated "To Public Service" on September 1, coinciding with Port Washington's centennial celebration.

The new plant was perhaps the largest structure in Wisconsin commenced and completed during the Depression. The project provided welcome paychecks for hundreds of construction workers and thousands of factory workers who supplied its machinery, including the turbine crews at Allis-Chalmers. The Port Washington plant was also a technological milestone. It used the same pulverized-coal system that had earned Lakeside a place in history, but TM's engineers added two new features: higher steam temperatures and "unit design." The Port plant was, in its final form, five power plants in one: Each unit had its own boiler, its own turbo-generator, its own smokestack, and its own transmission lines. The result was a new standard of operating efficiency. Although Lakeside remained the system's workhorse through World War II, the Port Washington plant immediately surpassed it as the world's most efficient, a title it held from 1935 to 1948. The powerhouse was also a welcome addition to Port Washington's landscape. It swelled the city's tax base by $7.5 million, assuming so much of the load that local homeowners were effectively exempt from property taxes for years.

Pushing Power

Although it had advantages that other industries did not enjoy, TMER&L did not remain solvent simply by coasting. S.B. Way and his fellow executives managed aggressively during the Depression, cutting costs, streamlining operations, and looking for avenues of growth. The industrial load, they soon realized, was hopeless; TM would see no gains there until the economy revived. Residential customers, on the other hand, particularly the unemployed, were spending more time in their homes, and they had shown, until the Depression, an increasing appetite for electrical conveniences. The system's managers attempted to revive and accelerate that momentum, working on two fronts at once. TM pushed appliance sales to build its residential load and, at the same time, stimulated consumption by lowering rates. The tangible result was a hell-for-leather merchandising campaign in the depths of the Depression.

Port Washington construction crews leveled a steep bluff (left) and used its soil to create acres of landfill for the new plant's coal pile.

Port Washington 100 Years Old | **MILWAUKEE SENTINEL** | Port Washington 100 Years Old

A TRIBUTE TO ENTERPRISE

Congratulations, Port Washington!

Celebration of the 100th anniversary of its founding by "The Port" calls for more than the ordinary felicitation.

Consequently the Milwaukee Sentinel in this special edition has attempted to depict in stories and pictures the Port Washington of today, the Port Washington of yesterday and its century of progress in between.

At the same time the Sentinel has recorded editorially and photographically the completion of the new Port Washington power plant by the Electric company, an achievement that coincides with the city's centennial.

Read this special edition, then go to Port Washington this weekend to participate in the city's biggest celebration.

The finished first unit became the world's most efficient power plant and a "tribute to enterprise" that was all too rare in the 1930s.

Merchandising had long been S.B. Way's strong suit. Power sales had been his first assignment in 1911 and, as president since 1925, he had launched any number of promotions. Way received expert assistance from another veteran whose star was on the rise: Gould W. Van Derzee. Born in upstate New York but raised in Milwaukee, Van Derzee was one of many TM managers with an engineering degree from the University of Wisconsin. After graduation in 1908, he taught at the Michigan School of Mines and then spent three years traveling for General Electric. In 1913 Van Derzee accepted an offer to come home as an assistant to S.B. Way, and he was soon following in his boss's footsteps. The young man took charge of the electrical side in 1918, earned a promotion to vice-president in 1925, and moved up to the general manager's post in 1934, the year of Way's sixtieth birthday. Gould Van Derzee was, in effect, the chief operating officer of the entire TM system during the Depression and war years. More deliberate and a good deal less autocratic than Way, he was widely popular among the system's employees.

S.B Way remained captain of the utility's team through the Depression ...

Way and Van Derzee tried everything in their efforts to build the residential load. In 1932 TMER&L began to invite garden clubs, church groups, and lodge auxiliaries to "spend an afternoon at the Electric Company." After a demonstration of "electric health cooking" in the Public Service Building's model kitchen, the ladies settled down to donuts, coffee cake, and bridge or canasta. Thousands more attended the annual Christmas show. Beginning in 1930, Home Service Bureau workers arranged a mouth-watering display of Christmas cookies, candies, and fruitcakes. Recipe books enabled local cooks to try their luck in their own (electric) ranges. Some TM missionaries went directly to the people. In 1933 the company began a round of demonstrations in neighborhood movie theaters. A complete meal was placed in an on-stage oven just before the main feature, and it was ready, pot roast and all, when the lights went on again.

Other appliances were promoted with equal vigor. In 1931 the PSB auditorium became an "ice cave," festooned with artificial icicles to boost refrigerator and freezer sales; 6,500 attended the show. Coin-operated "Meter-Ators" reflected the decade's emphasis on economy; consumers could pay for the small refrigerators a dime at a time, simply by dropping a coin into the convenient slot every day. NESCO roasters, made by the Milwaukee-born National Enameling & Stamping Company, were heavily promoted in the mid-1930s; NESCO sales in the metropolitan area soared from 5,500 in 1935 to 20,000 the next year. A new market opened up in 1935, when free light-bulb renewals were banned by state regulators. (The practice was considered "discriminatory.") The TM utilities were soon selling "Handy Bags" of six bulbs for one dollar — for use, they hoped, in new "Better Sight" lamps. The lighting promotions extended to commercial customers. In 1932 TM salesmen began to lug a "portable storefront" around to meetings of local merchants' groups. It demonstrated the "comparative lighting effects" of new fixtures that merchants could use to attract reluctant customers.

... with able assistance from Gould Van Derzee, a TM veteran who became general manager in 1934.

Way and Van Derzee did their best to whet the public's appetite for residential power. The Christmas Cooky display, which debuted in 1930, became a mouth-watering Milwaukee tradition.

Even small sales offices like the one in Cudahy offered a full range of electric appliances.

Free programs in the PSB auditorium attracted thousands of entertainment-hungry Milwaukee women.

In 1935 the system recruited a new worker to help with its load-building program: Reddy Kilowatt. The electrical elf with the light-bulb nose soon became a highly visible, and highly effective, ambassador of good will. In announcing his debut, *Rail & Wire* expressed confidence that Reddy would demystify the preternatural power of electricity:

> *He is not merely a symbol or trademark, but represents, in human fashion, the commonly used but little understood measure of electricity. Reddy Kilowatt is a living symbol.... He moves, he laughs, he works, he expresses emotion, he becomes thoroughly humanized.... No matter what position he is in, sitting, standing, walking, working he is always instantly recognized as Reddy Kilowatt — your electric servant.*

Reddy Kilowatt joined the TMER&L family in 1935, giving electric power a human face.

Most promotions of the 1930s targeted a single consumer group: women. Reddy Kilowatt's masters were well aware that women controlled residential demand, particularly in the years before power tools created a new generation of home handymen. Adopting the user-friendly little mascot was, at least in part, an attempt to win the trust of the region's housewives, and there were numerous others. In 1930 TMER&L launched *Fifteen Minutes at Home,* a twice-daily radio feature that offered tips on food preparation, tablesetting and, of course, appliance selection. In 1934 the company adopted "A Mixmaster for Mother" as its Mother's Day theme. A 1938 ad equated electric power with feminine freedom:

High-powered floodlights boosted sagging attendance at Borchert Field, home of the (minor-league) Milwaukee Brewers. Nearly 5,000 fans turned out for the first night game on June 6, 1935.

Less than twenty-five years ago, ... Mothers were supposed to be martyrs. It was inevitable. That Mothers should work "from sunrise to sundown" was commonly accepted, even by the women themselves, as their common fate. Then came electricity — the emancipator. Electrical adaptations in the home eliminated drudgery, lessened labor, conserved time.... A new freedom came into being. The "Martyr-Mother" of yesterday was replaced by the "Home-Manager" of today.

The campaigns worked, and they soon aroused the ire of competing retailers. In the World War I era, TMER&L and its sisters had been the dominant merchandisers in the region. As other merchants boarded the bandwagon (often beginning with radio sales), competition blossomed. When the consumer market began to shrink in the 1930s, other retailers cried that TMER&L's high visibility and ready access to customers gave the utility an unfair advantage, and they began to seek legislative redress. Recognizing that all appliances built load, regardless of who sold them, TM officials extended an olive branch to the independent retailers. The results were a cooperative advertising program, joint promotions, and a yearly convocation of dealers held in the PSB auditorium. As the "cooperating electrical fraternity" expanded its efforts, TM's role in direct sales diminished; by 1937 the company sold only 10 percent of the Milwaukee area's roasters and refrigerators.

Although its emphasis shifted from sales to support, the utility did not abandon the retail field. During a heat wave in 1936, TMER&L exhausted its entire inventory of 1,100 fans. In 1937 WG&E offered a group price on its "ensemble package," consisting of a flatiron, a waffle iron, a toaster, and a coffee maker. In 1938 TM launched perhaps its most unusual promotion: Every customer spending at least four dollars on an appliance received a two-and-a-half-pound package of cheese. More than seven tons of Wisconsin's finest were given away.

Electrical dealers and contractors filled the PSB auditorium for an Adequate Wiring sales rally in 1938.

The farmers who produced that cheese were another prime target of the load-building campaign. The TM companies, particularly Wisconsin Gas & Electric, had long been pioneers in rural electrification, and they mounted a new offensive in 1936. Line extension policies were liberalized, rates were cut, and farm families were encouraged to use the latest labor-saving devices in both their homes and their barns. The results were gratifying. By the end of 1937, the TM companies served the vast majority of farms in their service areas; TMER&L reached 82 percent, WG&E 70 percent, and Wisconsin Michigan Power 53 percent. All three companies exceeded the national average of 28 percent by a wide margin. The system's managers found that rural business was invariably good business; in 1932 TM's

Wiring the hinterland: Linemen connected a Wisconsin farmstead to company current in 1936.

Work groups like this crew in Fort Atkinson made the TMER&L system a national leader in rural electrification efforts.

directors noted that "rural customers were paying their bills much better than the city customers."

Aggressive sales efforts, both urban and rural, undoubtedly boosted demand for electricity; the ranges, refrigerators, and water heaters sold in 1936 alone added 12.6 million kilowatt-hours to the system's annual load, enough power to run several large factories for a year. But sales promotions represented only half the effort to increase consumption; rate reductions constituted the other half. Between 1929 and 1933, America's cost of living actually declined 24 percent, and deflation in one area of the household budget led consumers to expect deflation in all other areas, including their utility bills. Under significant pressure from both its customers and its regulators, the TM system cut rates repeatedly during the Depression.

Although they were to some extent making a virtue of necessity, TM's executives took an innovative approach to cost-cutting. Way and Van Derzee made rate reductions a strategic tool, and they squeezed every ounce of promotional value from every cut. In the spring of 1934, for instance, the "Free Kilowatt-Hour" program was unveiled. Residential, rural, and commercial lighting customers were encouraged to use unlimited power for the same price they had paid a year earlier. The two-month offer coincided with the spring run-off along the Wisconsin-Michigan border; most of the free electricity was "dump power" produced by hydro plants whose regular customers were practically out of business. Twenty-nine percent of those eligible took advan-

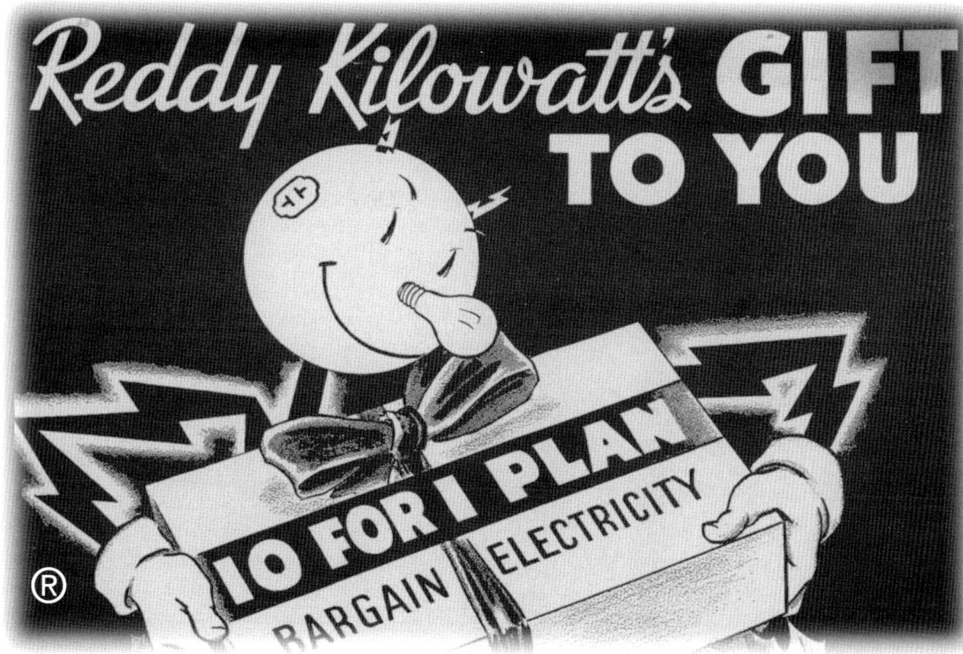

One of the utility's most popular promotions encouraged consumers to double their power usage for only 10 percent more than they had paid the year before. The 10 for 1 Plan helped TMER&L win the coveted Coffin Medal in 1938.

tage of the offer, and they used 50 percent more power than they had in the spring of 1933.

The free electricity program quickly evolved into the "10 for 1 Plan." Beginning in the spring of 1935, most customers could use up to 100 percent more power each month for only 10 percent more than they had paid in the same month of the preceding year. The company hoped that affordable power would increase the "electrical habit" among its patrons, generating enough volume to more than offset revenue losses. The 10 for 1 Plan ran continuously until 1945, and 90 percent of the system's customers used it at least occasionally.

As rates were reduced, they were also rationalized. In 1933 TM lowered rates for water heaters and other appliances that were used largely during off-peak hours. The entire residential pricing structure was modernized in the next two years. The TM companies still based some of their residential charges on the number of "active rooms" in a house, a measure that did not accurately reflect consumption. By 1935 all customers were charged on the basis of actual usage, with the cost per kilowatt-hour declining as usage increased. The new structure, incidentally, helped to end years of indelicate speculation about what occurred in an "active room."

The result of the various plans and adjustments was a dramatic drop in the price of energy. By 1939 TM's average residential customer paid only 2.9 cents per kilowatt-hour, down from 3.6 cents in 1935 and 4.5 cents in 1932. Just as S.B. Way and Gould Van Derzee had hoped, cheap power and vigorous merchandising combined to stimulate demand. In the depths of the Depression, residential energy use in TM's service area actually increased, running directly counter to the general trend of the economy. Annual power consumption per household rose from 599 kilowatt-hours in 1930 to 646 in 1932, 816 in 1935, and 1,004 in 1938. Fueled by residential demand, TM's annual output rose from a low point of 854.4 million kilowatt-hours in 1932 to a new record of 1.21 billion kilowatt-hours in 1936. Revenues recovered more slowly, the inevitable result of sweeping rate reductions, but TMER&L's performance earned the applause of its peers. In 1938 the company won the Coffin Medal, the electrical industry's highest award, for "the advancement of the electrical art" on several fronts, including the merchandising of power. The 1930s presented a variety of other challenges but, by the decade's end, it was clear that the "Electric Company" had successfully weathered the worst storm in its history.

A Watertown train barreled down the new Rapid Transit line near 22nd Street in December, 1930. The roadbed was later obliterated by Interstate 94. (Russell Schultz)

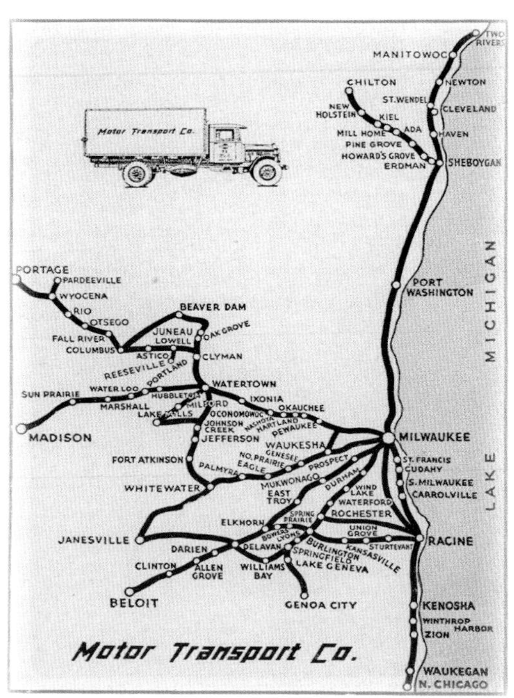

The Motor Transport Company, acquired in 1930, made TMER&L a regional power in the freight business.

A Troubled Empire

For managers of the electric utility, the challenge of the Thirties was to revive and intensify a demand that had already shown enormous potential for growth. TM's transit managers faced a radically different problem: to stem a drop in demand that had been developing since the World War I era. The decade began on a promising note. On September 22, 1930, the Rapid Transit line opened for passenger service west of Eighth Street. Although plans for the connecting subway from the Public Service Building were soon dropped, the new line put downtown Milwaukee within 30 minutes of downtown Waukesha. Its primary focus was interurban traffic, but Rapid Transit service attracted thousands of downtown commuters who lived on the city's West Side and in the western suburbs.

The high-speed line also served an ambitious new freight operation. In July, 1930, TMER&L purchased the Motor Transport Company, one of the largest trucking firms in southeastern Wisconsin. The combination of Motor Transport and Rapid Transit made TM a national pioneer in what is now termed "intermodal" transportation. Trucks and trains converged on the Transport Building, the new eight-story freight terminal at 10th and St. Paul, where containerized freight was transferred from one to the other for final delivery anywhere in the region. The company's goal, wrote transit manager Roy Pinkley, was "an efficient coordinated transport system using both facilities to best advantage."

The Transport Building was the freight system's nerve center. Like the Public Service Building before it, the

All freight lines converged on the Transport Building at 10th and St. Paul, where a maze of sidings and docks speeded the transfer of cargo from rail to truck and back again. (Russell Schultz)

new terminal was conceived as a multi-purpose structure, with abundant rental space on the upper floors. Some of it was used for storage and light manufacturing; General Electric's sales and service division was an early tenant, and the company's medical department, GE X-Ray, followed later in the decade. The new building also heightened the visibility of the transit system; a massive 1,212-bulb sign spelled out "Rapid Transit Freight Terminal" high above the Menomonee Valley.

Other projects continued the tone of general optimism. TMER&L budgeted $650,000 for track reconstruction on existing lines in 1930. A private right-of-way from Milwaukee to Racine was completed in the same year, extending Rapid Transit's reach to the south. Work on the Lakeside Belt Line began in 1930. The new line, completed in 1932, connected the Lakeside Power Plant with the East Troy interurban line at South 100th Street, and TM officials viewed it as an investment in the future. "The rapid development of the city toward the south," explained S.B. Way in 1930, "prompted immediate acquisition of a private right-of-way as the most economic means of making adequate provision for future transportation and electric service requirements."

The various projects started and completed between 1928 and 1932, from the downtown Rapid Transit line to the Lakeside Belt Line, required an expenditure of millions of dollars. Rapid Transit, wrote Roy Pinkley in 1930, "represents a huge outlay of money, but it is now established and

The Transport Building, pictured here in the 1940s, is now the home of the Aldrich Chemical Company.

will undoubtedly be of immense importance in ten years from the present time." He was dead wrong. The investments of the 1928-1932 period were, in hindsight, perhaps the most poorly timed in the company's history. As the economy soured, workers stopped working, shippers stopped shipping, and riders stopped riding. Automobile competition was no longer the major threat; vehicle registrations in Wisconsin actually fell 15 percent between 1929 and 1933. The problem was the economic privation of the riding public; as relief rolls swelled, people either walked or stayed home. TM's passenger traffic, which had been stagnant through the 1920s, dropped 4 percent in 1931 and 11 percent in 1932. Gross revenue declined even more sharply, falling 20 percent between 1931 and 1932 — double the decrease in electric revenue. The interurban lines operated at a loss, and no paying passenger ever boarded a car on the Lakeside line. Freight business was exceedingly scarce; the Motor Transport subsidiary did not come close to breaking even until 1934. Although no one expressed public regret that TM had invested so much in transit on the eve of a world-class disaster, there must have been some gnashing of teeth in the corporate offices.

The weekly pass gave riders in Milwaukee and Racine a powerful incentive to use public transit.

TM's transit managers tried desperately to stem the decline in ridership. Roy Pinkley could hardly sell appliances to stimulate demand, but he and his associates could cut fares, and they did so with uncommon vigor. The first promotion, and probably the most important, was the weekly pass, which offered unlimited riding for one dollar. First tried on the Racine lines back in 1919, the pass became a Milwaukee staple on May 4, 1930. TM raised the cash fare for a single ride from seven to 10 cents at the same time, and thrift-conscious Milwaukeeans were quick to choose the bargain. By the end of the program's first year, passholders made up fully 67 percent of the system's riders. The average cost per ride thus fell to 4.5 cents, less than the fare of 1890 for vastly superior service. The weekly pass had one unanticipated side effect: With less cash to handle, TM's motormen increased their average speed from 9.26 to 9.7 miles per hour in the first year.

Encouraged by the success of the weekly pass, Pinkley and his team launched a barrage of other promotions. During the 1930 Christmas season, TM unveiled a 75-cent shopper's pass good during off-peak hours. The pace accelerated in 1931, when the company introduced more liberal transfer privileges, a 75-cent shopper-theater pass, a 15-cent "pastime" pass on the city lines, and two interurban bargains: a 50-cent evening pass and weekend round-trip specials. Most of the promotions applied to non-rush hours; although commuter traffic continued to drop, off-peak business rose 9 percent in 1931. Two more inducements appeared in 1932: the familiar school pass (50 cents) and summertime "mystery tours" (25 cents) on the system's buses. (Only the drivers knew the destinations, which were usually parks or resorts outside Milwaukee.) By 1932 there were so many incentives to ride that the company began to publish *Transportation Bargain News*, a newspaper that was distributed free to 140,000 households.

TMER&L became a national leader in what it called "merchandised transportation." "Our company's part in the rebirth of the local transportation business is strikingly progressive," bragged *Rail & Wire* in August, 1931, and the nation's transit leaders agreed. TM won the coveted Coffin Medal in 1931 for its aggressive efforts to win back riders. True to form, the company immediately issued a 50-cent Coffin Award Pass, good for unlimited riding in non-rush hours.

TM's transit officials remained, for a time, optimistic. In the nation as a whole, public transit ridership fell 33 percent between 1929 and 1933. After three years of decline, passenger traffic on TM's lines rose 0.3 percent in 1933, the clear result of aggressive promotions. Company executives claimed that Milwaukee was outperforming every other major city but Washington, D.C., where hordes of New Dealers were keeping the cars and buses full. (The Washington system was another North American Company property.) "The trend is definitely upward," Roy Pinkley told the *Transit Journal* in October, 1933. "In fact, our business shows a disproportionately greater improvement than the average business in Milwaukee."

Two years later, Pinkley had changed his tune. Writing in *Rail & Wire* (December, 1935), he admitted that public transit was on the ropes:

> *For many years local passenger transportation in cities of this country has been acknowledged as a losing business. The regulations, restrictions, limitations and taxation burdens loaded on these transportation operations have made it exceedingly difficult to make ends meet financially.*

Pinkley might have added the riding public to his list of burdens, for the problem was, at root, one of demand. Although the various promotions helped to at least stabilize ridership, ridership and revenue are not the same thing. The merchandising campaigns never generated enough volume to offset the steep fare discounts. As a direct result, profit margins thinned to the vanishing point. In earlier years, the electric utility's earnings had more than made up for revenue shortfalls on the transit side, but the Depression ended that subsidy. The combined rate of return on both services fell from 7.79 percent in 1928 to 5.3 percent in 1932, and there was no longer enough profit to go around. In July, 1932, S.B. Way reported that electric revenues were "at a level which leaves no possible contribution by the electric utility toward the support of the railway service." It became obvious that the transit operation was dead weight that the electric side could no longer afford to carry. As the decade progressed, that weight became increasingly difficult to bear.

The transit system was finally forced to stand on its own, and its footing was anything but secure. As net income shrank, maintenance was deferred, purchases were postponed, and weeds were allowed to grow along the interurban rights-of-way. Government relief programs, ironically, added another layer of burdens. In an effort to put the unemployed back to work, the region's cities undertook extensive programs of street reconstruction, which required the removal of miles of streetcar track. Street reconstruction left TM's managers with two choices: to rebuild the rail lines, or to replace the streetcars with rubber-tired vehicles.

TMER&L chose rubber tires. In outlying areas of Milwaukee, motor buses took the place of trolleys. Bus routes had covered only half the mileage of streetcar routes in 1930, but they pulled even in 1937. (Because they dominated the high-volume downtown routes, streetcars continued to carry many more riders.) On

Rubber tires gradually replaced steel wheels as the foundation of TMER&L's transit service. The two modes met in 1934 at the transfer terminal on Murray and Farwell Avenues. (Irwin Scroggins)

the more heavily traveled crosstown routes, and in smaller cities, trackless trolleys were the substitute of choice. Equipped with rubber tires but tethered to double overhead wires, the trackless trolley was the ultimate hybrid; it combined the smooth ride and maneuverability of the bus with the power and roominess of the streetcar, and it was quieter than both. In 1932, following a street improvement project, Kenosha became the first city in America with an all-trackless electric transit system. Six trolleys were placed in service immediately, and the fleet soon grew to 22 vehicles. Milwaukee's first trackless trolleys appeared on North Avenue in 1936, shortly after the street was resurfaced. In the autumn of 1937, trackless service began on both the Mitchell-Forest Home and the 35th Street routes, and new trolleys replaced the streetcars on Holton Street in 1938.

Public reaction to the new vehicles was astonishing. Whatever magic the streetcar possessed had long since worn off, and TM's increasingly antiquated fleet inspired little devotion among its riders. Trackless trolleys, by contrast, offered both novelty and a superior ride. When service began on North Avenue, a local baker sent S.B. Way a cake with orange-and-cream frosting (TM's traditional running colors) and a note thanking him for "selecting North avenue as the No. 1 street in electric transportation." Ridership on the line increased 70 percent in the first year. When the Mitchell Street line opened, merchants used the occasion to stage a lavish celebration marking "A New Trail of Progress." The parade, speeches, street dances, and sales attracted 100,000 entertainment-starved Milwaukeeans, producing what TM's directors called "the largest gathering that ever congregated on a single street in Milwaukee."

The trackless trolley, a cross between the streetcar and the motor bus, revolutionized Kenosha's transit system in 1932. Kenosha was the first city in the country with an all-trackless fleet. (Russell Schultz)

Although company officials found the enthusiasm gratifying, they also found it troubling. Ridership on the new lines soared, but nearly all the growth was at the expense of nearby streetcar lines; riders were going out of their way to use the trackless trolleys. Even more disturbing was an obvious change in the public's tastes. As early as 1934, TM's managers were fielding citizens' requests to replace "the present cumbersome and noisy street cars" with trackless trolleys. As the requests piled up, the electric streetcar, long described as "the sturdy backbone of the transit system," became an endangered species, a situation with disastrous financial implications. In September, 1937, *Rail & Wire* published an upper-case lament:

> *The public, after being carried back and forth by electric railway cars for a half a century, is turning its back on the street railway car and is demanding rubber-tired transportation equipment. Owners of property along the street railway lines join the car riders in demanding this replacement by rubber-tired service, LONG BEFORE THE USEFUL LIFE OF RAILS, TROLLEYS AND RAIL CARS HAS BEEN USED UP.*

The Mitchell Street line was among the first Milwaukee routes upgraded to trackless-trolley service. Local merchants celebrated with an elaborate street festival. (Russell Schultz)

TM found itself between the proverbial rock and a hard place. The company was unwilling to invest in a thoroughly modern rail system, and equally unwilling to replace its old streetcars with a new fleet of rubber-tired trolleys. There was simply not enough profit potential in either option, which left the utility mired in a most uncomfortable middle. Even piecemeal replacement of old equipment soon exceeded the depreciation reserve, and TM fully expected things to get worse. "In view of the present trend of modern transportation," wrote S.B. Way in 1937, "the prospect of further abandonment losses ... becomes one of reasonable certainty."

The End of the Line

By the mid-1930s, at the latest, TMER&L was at a crossroads. The company had always been in the electrical business. Urban transit, like lighting and, later, power, was an electrical load, not an end in itself. Since World War I, however, the transit business had been turning away from the trolley tracks, and TM had been forced to follow it in directions Henry Villard could hardly have imagined. Buses, trucks, and filling stations had absolutely nothing to do with electricity, but all were part of the TM empire, and all were considered necessary to preserve the dominance of the electric streetcar.

Faced with mounting and multiple pressures — financial, public, and political — TM decided to get out. At some point between 1935 and 1937, the Electric Company decided to leave the transportation field. In the view of TM officials, the transit operation had become a dying limb that conditions of the 1930s forced them to amputate. The decision was, however, more easily made than executed; TMER&L was not a manufacturer who could simply close a division and move its assets elsewhere. Urban transit was a public necessity regulated by public officials, and TM had a legal obligation to maintain its service. It is no surprise, therefore, that the decision was never recorded in print, never put to a vote of the directors. The company decided to sever its transit ties slowly, step by step, in a process that ultimately took 15 years.

The first public step, and the most decisive, was taken in 1938, when TMER&L underwent a major corporate restructuring. Interest rates, like everything else, plummeted during the Depression, and TM wanted to replace its old debt with lower-cost obligations. Investment bankers, however, refused to refinance any company whose assets included a major transit system. (Failing systems had become a national phenomenon.) The solution was an internal divorce. On October 21, 1938, TMER&L absorbed the Wisconsin Electric Power Company, the paper corporation formed to hold the Lakeside Power Plant, and took that company's name as its own. The entire transit system was spun off into a new subsidiary at the same time: The Milwaukee Electric Railway and Transport Company, or TMER&T. The restructuring involved dozens of employees working 12-hour shifts for three months, and the pile of legal documents that resulted was more than two feet high. Twenty-five sets were signed by the company's chief executive. "I am glad I have a short name," said S.B. Way, "because the signing is part of my job."

The reorganization had several impacts. The new Wisconsin Electric Power Company (WEPCO) replaced its 5-percent principal mortgage with 3.5-percent debt, saving $850,000 in annual interest payments. With its transit arm severed, WEPCO also ensured its future access to favorable interest rates. Although it remained a Wisconsin Electric subsidiary, TMER&T, better known as the Transport Company, had its own financial statements and its own headquarters. Soon after the reorganization, Roy Pinkley and his corporate staff moved from the Public Service Building to the Transport Building. TMER&L, a name once synonymous with the best in urban transit, became a corporate ghost, a fading presence that still lingers above the doors of the old red-brick power plants and substations — and in the memories of a devoted group of railfans.

The new Transport Company came into existence with meager expectations. Its balance sheet included a $5.5 million reserve "for contingent losses on investment." Its retirement budget exceeded its capital budget by a wide margin. In the company's first three years, its financial officers made no provision for income taxes because, they wrote, "it is estimated that the Company has no taxable net income." The Transport Company proceeded to meet its managers' expectations. Although trackless trolleys continued to replace streetcars at a steady pace, the emphasis was on divestiture. The firm began to abandon or sell pieces of its system as fast as state regulators would allow.

The waiting room in the Public Service Building probably served more Greyhound Bus passengers than interurban train customers after 1940.

The interurban lines, long the most beleaguered members of the transit family, were the first to go. Service between St. Martins and Burlington ended in 1938. In sharp contrast to the delegation of dignitaries that had made the first run in 1909, arriving to band music and speeches, only one fare-paying passenger rode the last car to the end of the line. The East Troy branch was abandoned in 1939. (A portion of the line was sold to the Village of East Troy, whose citizens wanted to preserve their freight connection with the steam railroad at Mukwonago. In 1985 the East Troy Electric Railroad Museum took over operation of the railroad, and volunteers continue to operate it as the last electric line in the region.) Year by year, the interurban system shrank like a dying octopus. The Waukesha line abandoned service to Watertown in 1940, and the cars to Oconomowoc carried their last riders in 1941. On the old Milwaukee Northern line, the terminal was pulled back from Sheboygan to Port Washington in 1940.

The divestiture program affected every facet of the transit operation. Beginning in 1937, one year before TM's restructuring, a number of interurban bus lines were transferred to the Greyhound system, which moved its terminal to the Public Service Building in 1940. Three complete transit systems were sold soon after the Trans-

As public works projects unearthed miles of steel rails, the streetcar became an endangered species. A small crowd gathered to watch the work under way on 19th and Walnut Streets in 1939. (Irwin Scroggins)

port Company was incorporated: Racine's in 1939, Appleton's in 1941, and Kenosha's in 1942. (With the sales in Kenosha and Appleton, Wisconsin Gas & Electric and Wisconsin Michigan Power were completely out of the transit business.) The buyer in each case was a company headed by Henry Bruner, a transit entrepreneur whose career had begun in Indiana. The rail-truck freight operation, a source of disappointment through the 1930s, was the next to go. In 1939 TMER&T agreed to sell its Motor Transport subsidiary back to one of the trucking firm's founders. The deal was closed in 1943, and it included the Transport Building on St. Paul Avenue. (That building has housed a number of prominent companies over the years, including GE X-Ray, Motor Transport and, since 1969, Aldrich Chemical. It has also become one of the most unusually sited structures in the country: The old eight-story terminal is completely surrounded by the Marquette freeway interchange.)

Although it was shrinking rapidly, the Transport Company remained the dominant provider of transit service in Milwaukee and its suburbs. As the company's replacement program continued, the transit fleet's balance finally shifted from steel wheels to rubber tires; in 1941 TMER&T owned 493 trackless trolleys and buses, and 450 streetcars. (With fewer cars in service, the current consumed by the railway dipped from 142 million kilowatt-hours in 1929 to less than 100 million in 1941 — only 5 percent of WEPCO's output.) New equipment helped to bring in riders, but the company's financial reports, all of them featuring massive tax write-offs, showed less than stellar performance. In 1939, its first full year of corporate autonomy, the Transport Company reported net income of only $318,792. The transit subsidiary's payroll was substantially larger than Wisconsin Electric's, but its net income was less than 12 percent of its parent's.

Abandonments, replacements, and outright sales would continue for years, but the end was in sight by the late 1930s. The Transport Company was intent on going out of business, and nothing was going to stop it. Even World War II, which boosted ridership and revenues to record levels, was only a stay of execution. Wisconsin Electric decided that power and transit were no longer compatible, and the first integrated utility in the nation slowly disintegrated.

Strife and Strike

Managing a large utility during the Depression would have been difficult under any circumstances. In the case of TMER&L and its successor, Wisconsin Electric Power, the difficulties were compounded by the worst outbreak of political hostility since 1896. Public officials — local, state, and national — were determined to bring America's utilities to heel and, on the federal level, they were spectacularly successful. For S.B. Way and his management team, public controversy represented the third major front of the 1930s. During the same years that the company was building its residential load and backing away from its transit operation, executives spent a major portion of their time in a running battle with forces bent on controlling the utility's future. That battle raged both before and after the 1938 reorganization, and its impacts on the system were permanent.

Perhaps the most serious threat to TMER&L's autonomy arose from the most unexpected quarter: the utility's own employees. Since its inception in 1912, the Employes' Mutual Benefit Association had sustained one of the most comprehensive industrial welfare efforts in the region, embracing everything from athletics to education and from health care to collective bargaining. In its role as a "fraternal industrial union," the EMBA stopped considerably short of militancy, but the Association gave employees a voice that most felt to be sufficient — until the Depression.

Franklin D. Roosevelt encouraged a

Displaying the NRA's Blue Eagle signified agreement with New Deal labor standards.

more independent alternative. In June, 1933, barely three months after his inauguration, Roosevelt signed the National Industrial Recovery Act into law. The NIRA relaxed antitrust rules and other restrictions, making it easier for businesses to do business, but, in a conscious effort to maintain a balance of power, it also opened new doors to organized labor. The National Recovery Administration (NRA), which administered the new law, enforced the right of employees "to bargain collectively through representatives of their own choosing ... free from the interference, restraint, or coercion of employers." S.B Way signed the NRA's "Re-employment Agreement" in September, 1933. He did so with some reluctance but, as the head of a public utility, Way felt it prudent to "cooperate for recovery." TMER&L was soon displaying the NRA's Blue Eagle emblem as a tangible sign of its accord with federal authorities.

Hoping to preempt outside organizers, hundreds of employers scrambled to establish company unions that would, they hoped, meet NRA guidelines. TMER&L, of course, already had one of its own, and S.B. Way was confident that the EMBA would remain the sole voice of his system's employees. Way's confidence was misplaced. Some workers clearly wanted representatives who were less closely allied with management, and labor organizers, who derided company unions as "kiss-me clubs," developed an intense interest in a firm employing 4,700 people. In the autumn of 1933, as the ink on the NRA agreement was drying, three union locals were organized among TM's workers, one for trolley and bus drivers, another for electrical workers, and a third for power plant employees. All were affiliated with the American Federation of Labor. (The transit local was by far the largest and most vocal, providing management with one more reason to reconsider its role in the transportation business.) The unions represented a minority of TM's employees — 200 by management's count and 1,400 by the AFL's — but they also represented an abrupt shift in the company's labor relations.

S.B. Way proved absolutely intransigent in dealing with the unions. He blasted the "professional labor agitators" who were stirring up trouble, and in 1934 he fired eight union members who were seeking recruits among their co-workers. That, it turned out, was a major strategic blunder. When the unions complained, federal authorities stripped TMER&L of its Blue Eagle. Despite this highly public rebuke, Way refused to rehire the activists, even after the EMBA urged him to do so. "The company," he fumed, "is being

Thousands of Milwaukeeans took to the streets during the 1934 strike, including this belligerent crowd at the Kinnickinnic Avenue car station. Fifteen people were arrested and 10 were hurt before the evening ended. (Milwaukee Journal Sentinel)

subjected to unjustifiable persecution, abuse and vilification." Faced with top management's adamant refusal to negotiate, the unions called a strike for June 26, 1934. It proved to be the most serious in a local epidemic of post-NRA work stoppages; the number of strikes in Milwaukee soared from six in 1933 to 42 in 1934.

After weeks of general tension and front-page news coverage, the actual strike was, at first, almost anticlimactic. A small minority of workers — 157 by TM's count and 400 by the AFL's — failed to report for work on June 26, and they walked the picket lines in a demoralizing rain. Soon, however, the general public made the strikers' cause their own. TMER&L had never completely shed its image as a heartless monopoly, and Way's apparent arrogance made the company a lightning rod for the general disaffection of the unemployed. Thousands gathered at the Kinnickinnic car barns on the first night of the strike. "It's the people of Milwaukee against the chiselers and the big guys," declared one demonstrator. "If they lose this strike, unionism is dead in this town." Like their parents and grandparents in 1896, the crowd expressed its displeasure physically, pelting the streetcars with bricks and rotten eggs. (TM had "armored" its cars with wire mesh, and the "bird cages" were attractive targets.)

On the evening of the second day, June 27, 24,000 demonstrators laid siege to the system's car stations. They were met with fire hoses at Fond du Lac Avenue and tear gas at the Kinnickinnic barn. Rioting became so general that streetcar service ground to a halt, practically paralyzing the city. Concerned about the adequacy of police protection, S.B. Way served notice on Mayor Daniel Hoan that TMER&L would hold the City of Milwaukee liable for any damage to its property. Hoan shot back a stinging reply:

The 'Strike'

Our company and ninety-six per cent of the employes believe that the public is primarily interested in adequate and continuous public utility services. This belief is demonstrated by our labor contract of 1918 providing for arbitration as the means of settling disputes.

In our opinion, the "strike" leaders never hoped for more than publicity out of the present agitation. They expected that federal authority would step in and "stop" the "strike".

If strike agitators again threaten the safe and convenience of well over half a milli people, the public can count on the contin service of our employes and their keen ap ciation of their obligation to the public.

The Service Will Not

S. B. Wa
President,

THE MILWAUKEE ELECTRIC RAILWAY AND LIGHT COMPANY
Service First

DON'T SCAB

SUPPORT YOUR CLASS BROTHERS IN THE FIGHT AGAINST THE ELECTRIC TRUST

To All users of Electricity, Street Cars and Buses:

Do not ride the street cars or buses—don't allow yourselves to be caged like animals for the profits of J. P. Morgan & Co., the real owners of the T.M.E.R. & L. Co.

The street cars are not safe. In case of accident you will be unable to escape. You will be trapped in these cages.

The workers of the Electric Co. are on strike for higher wages and the right to organize. If these workers win it will mean better conditions for all of us. It will be a smashing blow against the company union.

Militant picketing will stop the street cars from moving.

Join the picket line!

Make the picket line so strong that not a car will move!

Go to the Electric Co. in a body and demand your money back for the weekly pass!

Refuse to ride in scab cages!

Demand the withdrawal of the police from the picket line!

Raise demands for higher wages and better conditions in your shop.—Take up the question of sympathetic strike!

Unite behind the Electric workers!

Issued by the Communist Party, District No. 18—617 North 2nd St.—Room 510
Milwaukee, Wisconsin

A leaflet war between union sympathizers and TMER&L management accompanied the picket lines and protests.

With one protester dead and a blackout imminent, the company threw in the towel. Sealing a new labor agreement, S.B. Way shook hands with Edward McMorrow, an official of the international transit union, while mediator Julius Heil stood between them approvingly. (Milwaukee Journal Sentinel)

I now notify you ... that you alone are solely responsible for the riots that have so far blotched the good name of the city.... Your attitude toward your employees, our people, our city, our federal government, is more arrogant than that of any ruler in the world. Not since the days of King George III of England has any such ruler successfully defied our nation. But you impudently refuse to comply with the reasonable request of the representatives of the United States Government until Uncle Sam himself has been compelled to rebuke the insolence by removing your Blue Eagle. You are now witnessing the harvest of pent-up public indignation you yourself have aroused.

Strong words, and they did nothing to quell public passions. The strike built to a climax on the evening of June 28, when 30,000 Milwaukeeans took to the streets. Among them were 5,000 who marched on the very source of TM's power: the Lakeside plant. Chain-link fences and imported guards were not enough to stop a phalanx of young men intent on storming the powerhouse. One of them reached the plant and apparently thrust an iron pipe through a ground-floor window, making contact with a live circuit. He was killed instantly.

Everyone involved suddenly realized the magnitude of the stakes they were playing for. The crowd dispersed, slowly but quietly, and a majority of Lakeside's workers, apparently fearful for their own lives, decided not to report for work the next day. On the morning of June 29, a sober S.B. Way finally sat down to negotiate with the unions. He had little choice; the strike had already caused one death and 50 injuries, and the Lakeside plant was on the verge of a shutdown that would have left most of the region in darkness. Moderated by industrialist Julius Heil (who had all parties recite the Lord's Prayer before getting down to business), the marathon bargaining session produced an agreement that was a total victory for the unions. The fired workers were rehired, all strikers were reinstated with no loss of seniority, and the EMBA was reduced to a fraternal benefit society. The three trade unions easily won election as bargaining agents for their members, and 90 percent of TM's clerical and sales workers voted to join a fourth union later in 1934. The utility has been a union company ever since.

S.B. Way accepted the June 29 settlement with public grace, announcing that it had been made "in the interests of public peace and order." In the quiet confines of the TM boardroom, however, he complained that the terms had been "forced" on him "under coercion and duress." Way's duress was far from over. Even before the strike, TMER&L heard familiar rumblings from City Hall. In the changed political climate of the 1930s, Dan Hoan revived his long-term campaign for municipal control of electric power, and he had the support of a Common Council majority. In December, 1933, Milwaukee's aldermen voted to submit a $15 million bond issue to referendum. The proceeds were to have funded a City-owned power plant on Jones Island, which would have operated in direct competition with TMER&L. The referendum question was withdrawn in early 1934, when Milwaukee applied for a federal public works loan to begin construction. The application was denied, leading local officials to try a third tack in 1935: buying all of TM's electric properties within the city limits. The question was scheduled for a referendum on April 7, 1936. (It covered, interestingly, only those properties "used and useful for the distribution of electric light and power," a sure sign of the transit system's decline.)

Municipal ownership referenda were nearly as common as strikes in the mid-1930s. The question was so popular that it became the national high-school debate topic for 1936: "Resolved: That all electric utilities should be governmentally owned and operated." Between 1934 and 1936, voters in 342 communities around the nation considered the question, and municipal advocates won in more than half the cases. TM officials had ample cause for concern. In the weeks before Milwaukee's referendum, they mounted a full-scale campaign against the proposal, charging that it was "indefinite in its terms and unfair to the electric users and taxpayers of Milwaukee." (Although figures of $35 million to $60 million were bandied about, the question included no cost estimates.) TM argued further that cutting off a regional system at the city line would amount to a "dismember-

With capitalism under siege, Mayor Dan Hoan stepped up his campaign for municipal ownership of the electric utility. (Milwaukee Public Library)

The company countered that municipal ownership would lock the citizens of Milwaukee in a painful "tax trap."

ment ... destructive of the interests of both the public and the Company."

Voters apparently agreed. On April 7, they turned down the proposal by a 4-3 margin — a considerably slimmer margin of victory for TM than the 2-1 rejection of a similar referendum in 1925. Dan Hoan was livid. "The big banker-controlled Electric Company and its allies," he fulminated, "through the use of unlimited funds, employing the mails, the press and all the opposition candidates, successfully hoodwinked Milwaukee." No one knew it, least of all Dan Hoan, but the 1936 referendum would prove to be the last hurrah for municipal ownership in Milwaukee.

Local strife had its counterpart on the state level. In 1930 Philip La Follette, the younger son of "Fighting Bob," won the governor's seat his father had held a quarter-century earlier. In the true Progressive tradition, La Follette had made "the need to limit the concentration of economic power" his principal campaign issue, and the "power trust" was one of his favorite targets. The 1931 state legislature passed a number of La Follette-backed bills inimical to Wisconsin's private utilities, including several that made it easier for local and state governments to enter the electric business. Most had little impact on TMER&L, but the legislature also addressed issues of regulation. In 1931 the old Railroad Commission was rechristened the Public Service Commission, a name more in keeping with its responsibilities, and the PSC's power over the financial practices of state utilities was broadened significantly.

The Death Sentence

Political hostility on the local and state levels was nothing in comparison with national developments. Even before the Depression, many Americans had found the headlong growth of utility holding companies profoundly troubling. Too much power, they felt, both electrical and political, was concentrated in the hands of a few mega-utilities, and Samuel Insull's empire was nearly always Exhibit A. By 1929 Insull, through Middle West Utilities and a pyramid of related firms, controlled roughly one-eighth of the nation's power supply, reaching 20 million people in 32 states. Operating subsidiaries were regulated by individual state commissions, but the holding companies themselves were, in effect, above the law.

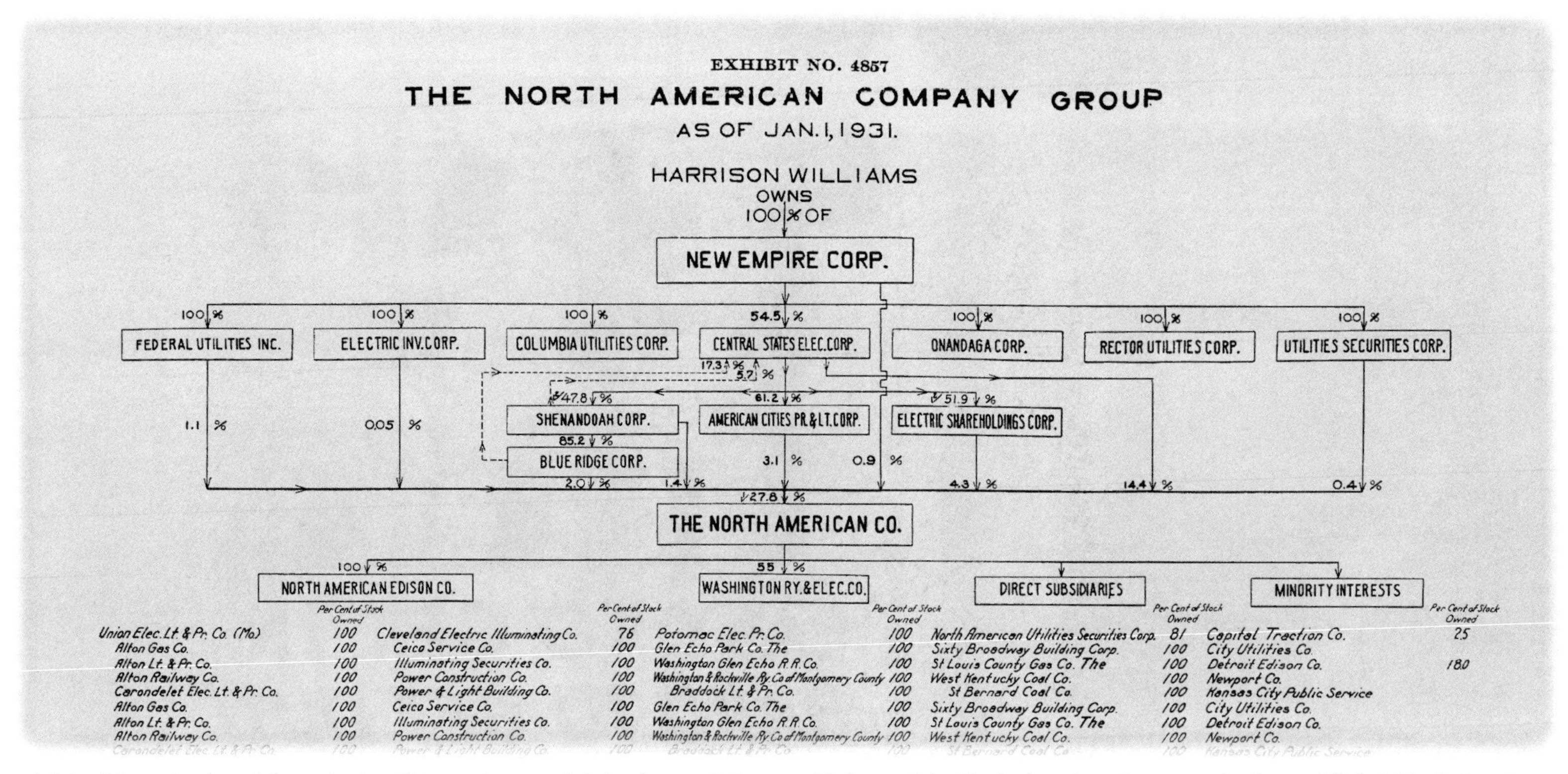

A federal investigation of the nation's utility empires revealed the characteristic pyramid shape of the North American Company. It also established North American as one of the best-managed companies in the industry.

The complete absence of national regulation seemed to call for action, even during the "business decade" of the Twenties. In 1928 the Federal Trade Commission launched a full-scale investigation of America's holding companies. The FTC found that some firms, including Insull's, indulged in practices that ranged from the unsound to the piratical. Operating subsidiaries were charged exorbitant fees for engineering and financial services, properties were transferred from one level of the pyramid to another at inflated prices, and outside shareholders were deliberately kept in the dark. The disclosures led Franklin Roosevelt to denounce "the Insull monstrosity" as "a kind of private empire within the nation."

The North American Company, which approached Middle West Utilities in size, emerged from the FTC investigation as a white knight, relatively speaking. Its accounting policies were conservative, its central staff was lean (177 people in 1930), and its services to subsidiaries were provided at actual cost. Edwin Gruhl, the Milwaukee native who became North American's president in 1932, testified that the loose-knit federal style of Henry Villard's era still governed relations with subsidiaries:

> *Autonomy prevails among the various public-utility companies under North American ownership. At the heads of the various companies ... are men ... fully familiar with the special conditions and problems local to the territories under their management. Upon them is imposed the necessity for initiative, self-reliance, and independence of thought and action.*

Samuel Insull, pictured on his way out of jail in 1934, was blasted by Franklin Roosevelt as the head of a "private empire" that threatened the foundations of American democracy.

The Depression was under way by the time the first FTC reports were released, and the holding companies found themselves in economic as well as political hot water. Every pyramid rested on a foundation of operating subsidiaries like TMER&L. As sources of operating income dried up in the 1930s, the stream of profits rising toward the top of the pyramid slowed to a trickle and, in many cases, stopped completely. Harrison Williams, North American's principal stockholder, reportedly watched his assets plummet from $612 million in 1929 to a mere $5 million in 1934. The tycoon was ultimately forced to sell four of his five homes and to cut his domestic staff from 25 servants to 10. But Williams' travails paled in comparison with Samuel Insull's. In 1932 the House of Insull collapsed in spectacular fashion, burying thousands of small investors in the rubble. The story of his subsequent flight, extradition, and trial shaped the subsequent course of public discussion. Although the aging patriarch was acquitted on all charges, the failure of America's largest holding company seemed to confirm the most extreme charges of the critics.

A few voices were raised in defense of the prevailing system. In his testimony before the FTC, Edwin Gruhl argued that holding companies offered economies of scale that worked to the advantage of local utilities:

It is by no means impossible, or even improbable, that the process of integration in groups which has gone on under the corporate control of the North American Company would have been brought about had each of the major companies been under independent control. It is obvious, however, that the intervention of the holding company has been of tremendous value in accelerating and facilitating each of these entirely logical and natural group developments.

The industry's more responsible voices were lost in a chorus of catcalls. Gruhl's own voice was stilled by a fatal heart attack on January 21, 1933, shortly after his forty-sixth birthday. Frank Dame came out of retirement to lead North American until his own death a year later; in 1934 the reins passed to James Fogarty, who had begun his North American career as a stenographer in 1902.

When Franklin Roosevelt entered the White House, he found the mood ripe for punitive action against the holding companies. Other utility initiatives, including the Tennessee Valley Authority and the Rural Electrification Administration, took precedence, but in 1935 Congress passed Roosevelt's Public Utility Holding Company Bill. The new law's most potent feature was its infamous "death sentence" clause: Every holding company was required to pare its operations to a single integrated utility system by January 1, 1938. The law's ultimate goal, in other words, was abolition of the holding company empires, a point Roosevelt had made clear in his message to Congress:

Except where it is absolutely necessary to the continued functioning of a geographically integrated operating utility system, the utility holding company with its present powers must go.... It is a corporate invention which can give a few corporate insiders unwarranted and intolerable powers over other people's money. In its destruction of local control and its substitution of absentee management, it has built up in the public-utility field what has justly been called a system of private socialism which is inimical to the welfare of a free people.

Roosevelt was out to slay a dragon, but he tried to reassure the industry's white knights. "The good holding companies," he said, "such as North American, have nothing to fear from the Holding Company Act." It was an empty promise; in the end, all holding companies were felled by the same lance. An interminable series of court challenges, including some led by North American, delayed enforcement for years, but the Holding Company Act cleared every legal hurdle. Slowly, by degrees, North American and its subsidiaries began to comply. The 1938 restructuring that turned TMER&L into Wisconsin Electric Power was, at least in part, an attempt to consolidate local operations. Late in the same year, North American filed its first "integration plan" with the new Securities and Exchange Commission. Although Wisconsin Electric was the firstborn, it was no longer the parent company's favorite child. North American proposed keeping its St. Louis and Cleveland properties and divesting itself of the Milwaukee and Washington, D.C. systems. In 1941, preparing for the end, Wisconsin Electric acquired the common stock of both Wisconsin Gas & Electric and Wisconsin Michigan Power; its "sister companies" thereby became direct subsidiaries in a single integrated utility system.

The dissolution of the holding company empires rivaled in scale the court-ordered breakup of AT&T in 1982. Not even World War II was enough to slow it down. SEC officials began to carry out the "death sentence" in 1940, and their efforts continued for more than a decade. In 1942 North American was informed that "a single integrated system" meant just that. When the Supreme Court refused to hear its case, the company dropped its plans to retain the Cleveland utility and began to transfer all its eggs to the St. Louis basket. In 1947 North American distributed the common stock in its other operating subsidiaries, including Wisconsin Electric, to their individual shareholders. In 1954, with no rationale for its continued existence, America's first utility holding company, and long one of its largest, slipped quietly beneath the waves, leaving the St. Louis utility in the hands of its investors.

The change in Wisconsin Electric's status, from a wholly owned subsidiary to an independent utility, was both radical and, from the public's viewpoint, practically invisible. The company's management staff, its operating policies, and its internal culture changed not at all. It is likely that most customers and many employees were only dimly aware of the shift in ownership. WEPCO lost the ample legal and financial resources of its parent, but the utility was hardly orphaned. After years of functional independence, Wisconsin Electric was long past its adolescence and ready for freedom.

Although its day-to-day impacts were minimal, the breakup of North American certainly marked the end of an era. In its relentless campaign to

Dan Hoan's 24-year tenure in the mayor's office came to an end in 1940, when he was unseated by upstart Carl Zeidler. (Milwaukee Public Library)

rein in runaway capitalism, the Roosevelt administration had made WEPCO a purely local enterprise. No longer could critics like Dan Hoan charge that "the largest trust in the world" had "fastened [its] hands ... upon our pocketbooks." In 1940, as if to confirm the era's closing, Daniel Webster Hoan was forced to leave City Hall, after a tenure of 24 years. He was narrowly defeated by Carl Zeidler, a vigorous young city attorney with an abundance of charm and a near-total absence of ideology. Zeidler quickly established a cordial relationship with the utility. The first of Milwaukee's singing mayors, he helped to dedicate a series of new trackless-trolley lines by leading his constituents in song. Fifty years of political controversy came to a melodious end, and Wisconsin Electric's managers returned to the business of managing.

The new mayor joined the Trainmen's Quartet in song when the Center Street trackless-trolley line opened in December, 1941. Zeidler went off to war a few months later, never to return. (Milwaukee Public Library)

Waking to War

In the last years of the 1930s, as utility lawyers were testing the Holding Company Act in countless courtrooms, a much larger development was taking place: economic recovery. Order rates were on the rise, relief rolls were shrinking, and the high tide of unemployment was finally going out. Wisconsin Electric and its affiliates felt the return of good times directly. The system's annual output, which had been growing steadily since 1933, jumped to 1.6 billion kilowatt-hours in 1936 and approached the 2-billion mark in 1940. Revenues grew more slowly, reflecting the steep discounts and multiple incentives of the early Thirties, but the worst years were over. The 10-percent pay cut was restored in 1937, and common stock dividends were resumed in 1939. (Except for a brief hiatus at the start of World War II, quarterly dividends have continued ever since — one of the longest streaks in the American utility industry.)

As demand rose to new levels every year, the system's engineers decided that they needed more capacity. In 1937 the utility began work on major

New units at the East Wells (left) and Commerce Street Power Plants helped Wisconsin Electric meet growing demand during a major military build-up.

additions to its two oldest powerhouses: East Wells and Commerce Street. The East Wells "boiler house" was completed in early 1939, and Commerce Street's followed in late 1941. Although the new units were designed to serve the steam heating system in downtown Milwaukee, they added nearly 50,000 kilowatts (roughly 10 percent) to WEPCO's capacity. The engineers launched an even larger project before the Commerce addition was finished. On September 17, 1940, company executives broke ground for the second 80,000-kilowatt unit (of five planned) at Port Washington.

There was comparable expansion along the state's northern border. In 1937 Wisconsin Michigan Power bought two hydro plants, Chalk Hill and White Rapids, on the Menominee River, adding nearly 16,000 kilowatts to its 19,630-kilowatt capacity. In 1941 the company built a dam of its own on the Michigamme River, creating a reservoir that stabilized the flow of water to hydro plants on the Menominee. (Designated the S.B. Way Dam, it remains the only structure in the WEPCO system named for an individual.) In October, 1941, Wisconsin Michigan crews began construction of the 15,000-kilowatt Peavy Falls plant, 12 miles downstream from the Way Dam. Completed a year later, it was the largest hydro facility in the Wisconsin Electric system.

It was not the return of "normal" conditions that sparked the various building projects and boosted the demand for power. With the rise of the Nazis in Germany and militarists in Japan, the world was increasingly on

The Peavy Falls hydro plant, completed in 1942, added capacity in Wisconsin Michigan Power's territory.

Milwaukee's contribution to the war effort included these bomb casings produced at the A.O. Smith plant. (Milwaukee Public Library)

As demand soared, WEPCO rushed the second unit of its Port Washington plant to completion in 1943.

edge, and the United States looked to its defenses. A military buildup was under way well before the outbreak of European war in 1939, and it gathered momentum through the decade's end. A sense of impending crisis gripped the nation. In 1939, at the War Department's urging, Wisconsin Electric closed its power plants to visitors, hoping to prevent sabotage. In 1940, fearful of impending shortages, WEPCO ordered a turbine for its Port Washington plant and a year's supply of tires for its transit fleet. S.B. Way announced the Port plant addition as his company's contribution to "power preparedness for national defense."

With the bombing of Pearl Harbor on December 7, 1941, defense production shifted into overdrive. The nation moved directly from an economic emergency to a military emergency.

World War II placed an enormous burden on Wisconsin Electric. From the iron mines of Upper Michigan to the factories of Milwaukee, Racine, and Kenosha, the system's customers played a pivotal role in the wartime production effort, and their dependence on electricity was absolute. "The great industrial area which we serve has become a mighty arsenal," the company declared in its 1941 report. "The job of providing an adequate and dependable supply of electric power for the mighty defense effort in this great area falls upon us." Wisconsin Electric was up to the task, particularly after Port Washington's second unit went on line in 1943. The system's combined output more than doubled between 1938 and 1944, rising from 1.7 billion kilowatt-hours to 3.44 billion, and heavy industries accounted for most of the growth. Sales to industrial customers climbed from 34 percent of the utility's output in 1932 to 52 percent in 1943. Continuing rate reductions and stiff wartime taxes kept profits below 1929 levels, but Wisconsin Electric was clearly doing its part.

The war also breathed new life into the transit system. As automobile production was suspended and gasoline and tires were rationed, the

Wartime shortages led to crowded streetcars and crowds of pedestrians. This was the scene at Third and Wisconsin in December, 1944. (Irwin Scroggins)

growing legions of war workers depended on public transit. With most plants working three shifts, the trolleys and buses were busy day and night. Passenger traffic on the old TM lines skyrocketed from 218 million riders in 1938 to 428 million in 1944 — an all-time record. The Transport Company tried valiantly to distribute demand. Stops were spaced more widely, charter business was discontinued, and "women at home" were asked to "Ride More from 9 to 4." In 1942 Milwaukee's employers adopted a system of staggered working hours to relieve pressure on the transit lines.

The problems were aggravated by shortages of both manpower and vehicles. With the armed forces and defense plants siphoning off workers, drivers and motormen were in short supply, but the company was able to increase its work force from 2,486 in 1938 to 2,869 in 1943. ("Be a Transport Company 'Pilot,'" urged one ad.) Rolling stock was a good deal harder to come by. Fortunately, transit officials had ordered 50 trackless trolleys before a wartime moratorium was imposed; the system's 1942 fleet consisted of 274 trackless trolleys, 293 buses, and 440 streetcars. The new trackless lines offered up-to-date service, but some of the streetcars that had been pressed into service were practically museum pieces.

As traffic on the old routes rose to record levels, the company opened one new line: a 0.6-mile stub serving the Allis-Chalmers supercharger plant in what is now West Milwaukee. It was the very last addition to the rail system. Despite runaway growth on the transit side, all concerned knew that the boom was temporary, and the decision to leave the field was never reconsidered.

Helping to win the war at home, Wisconsin Electric used its streetcars to promote War Bond sales. (Don Leistikow)

The transit subsidiary used its wartime profits to retire stocks and bonds held by Wisconsin Electric, a likely prelude to a sale of the entire system, and the divestiture program continued through the war years. Between 1942 and 1945, major components of the old TMER&L transit empire were sold off: the Kenosha transit system, the Motor Transport Company, and the interurban rail lines connecting Milwaukee with Racine, Kenosha, and Port Washington. All but Motor Transport were sold to Henry Bruner's Kenosha Motor Coach Lines. By war's end, the Waukesha and Hales Corners routes were all that remained of the once-proud interurban system. They, too, came under Bruner's control in 1946.

For the people of the WEPCO system, whether they worked on the electrical or the transit side, the demands of wartime posed an unusual challenge. In comparison with the swollen payrolls of local defense plants, growth in the work force was modest; the Electric Company and the Transport Company together employed 5,300 people in 1943 — a significant increase from the low point of 4,600 in 1933, but substantially below the 1929 peak of 6,500. The system was distributing more power and carrying more riders with fewer people than ever before.

What kept their productivity high was a commitment to the war effort. A total of 1,014 system employees joined the armed forces. (Sixteen did not return.) Those on the home front had ample opportunity to display their own patriotism. More than 4,500 workers earned Civil Defense certification in everything from air raid procedures to incendiary bomb identification. Hundreds helped to guard WEPCO's power plants as armed "auxiliary MPs." Every War Bond drive met or exceeded its quota. Produce from more than 400 Victory Gardens was displayed in the Public Service Building's auditorium. Warehouses and stockrooms were scoured for usable scrap; in 1942 Wisconsin Electric turned in more than 3,000 tons of scrap metal for recycling. Trackless trolleys were billed as "transports for production soldiers" and powerhouse smokestacks as "cannons of industry." In 1944, with perhaps a touch of hyperbole, *Rail & Wire* congratulated its readers on their patriotic spirit:

> *Hitler blundered tragically. He didn't reckon on the matter of electric and transportation utility workers. He didn't count on you as war workers on the home front providing services so essential to our amazing war production program.*

WEPCO tried just as hard to keep its customers involved in the war at home. Two streetcars became red-white-and-blue billboards for War Bond drives. The PSB auditorium, which was remodeled early in the war, provided a venue for draftee orientations, production award ceremonies, USO dances, and other patriotic gatherings. The Home Service Bureau stayed in business, but its emphasis shifted from consumption to conservation. Classes were offered in canning, freezing, appliance repair, cooking with soybeans, and "preparation of lunches for war workers." All housewives were urged to take the Homemaker's Pledge:

> *I will conserve electricity and everything I now own and use. I will take good care of my appliances. I will use them efficiently and wisely. I will keep them in good repair. By so doing, I will eliminate waste of time, money and effort, and thereby contribute materially to the all-out war effort.*

The emphasis on appliances reflected a 1942 ban on new labor-saving devices. WEPCO's appliance repairmen soon reported "exceedingly heavy" demand for their services, and in 1943 the company opened a used-appliance exchange for its customers; 9,600 devices changed hands in the first year. Farmers were a notable exception to the ban. Agriculture was a defense-vital industry and, with hired help in short supply, farmers turned increasingly to "wired help." Defense planners encouraged more liberal line-extension policies and more general use of the latest electrical equipment. In 1944 alone, farmers in Wisconsin Gas & Electric territory purchased 1,582 milking machines, 1,500 water pumps, 673 poultry brooders, and 143 milk coolers. (*Rail & Wire* observed in 1944 that the windmill, once "as much a farm landmark as the barn or the silo," was fast disappearing "because of the widespread use of electric water systems.")

The ration books and the shortages, the controls and the quotas were finally phased out after the surrender of Japan on August 15, 1945. In four years of actual war and several more of preparation, the Wisconsin Electric system had supplied dependable power and transit service to a region that was a linchpin in America's defense effort. The system's employees

USO dances attracted crowds of servicemen to the Public Service Building auditorium.

felt a justifiable sense of accomplishment. "We recall with some pride," wrote S.B. Way shortly after V-J Day, "that the Company ... was able to meet unprecedented demands for electricity for war production without curtailing its supply for essential civilian purposes." Like virtually all of his fellow citizens, Way also looked ahead to better times: "We share in rejoicing over our country's victory and the redirection of human effort to constructive pursuits of peace."

The years between 1930 and 1945 were eventful by any standard. The Wisconsin Electric system weathered a financial catastrophe, a bitter strike, rabid political antagonism, the imminent demise of its parent, and a war emergency, and it did so without losing its composure or its sense of direction. The utility, in fact, emerged from the period with a new name, a new corporate structure, and a final solution to its transit problems.

However eventful, however painful at times, the 1930-1945 period was, in a long view of the century, an interruption. Some developments, particularly the radical expansion of federal authority, proved to be permanent, but it was the preexisting patterns that showed real staying power. Trends of the Roaring Twenties — new consumer markets, the mass media, suburban growth, and dizzying technological progress — had come to a temporary halt, only to emerge, 15 years later, with abundant new energy. The nation was on the verge of explosive growth, and Wisconsin Electric was poised to grow with it.

Rising above the rigors of depression and war, the workers of WEPCO began to splice together a new future for themselves and their company.

Chapter 7

1945-1965

Golden Years

With the coming of peacetime, America's electric utilities entered a golden age that lasted for at least 20 years. Gone were the turf wars of the Twenties, the economic strain and the political strife of the Thirties, and the multiple pressures of World War II. Pent-up consumer demand pushed the economy into overdrive, and utilities multiplied their output to keep pace. Never before had electric power played such a pivotal role in the life of the nation, and never since have there been so few impediments to the industry's growth.

At the time, of course, very few industry leaders felt that they had entered the Promised Land. Memories of the Depression were still fresh, and some feared that the economy would collapse again without the stimulus of war. There were, in fact, a few bumps in the road after 1945: a bout of postwar inflation, periodic recessions, and the numbing anxieties of the Cold War. But the ride was unusually smooth. In comparison with what came both before and after, the years between 1945 and 1965 were a period of almost preternatural calm. The story of Wisconsin Electric after the war is, in its essentials, little more than the chronicle of a successful, well-managed utility going about the business of spectacular growth. There were no traumatic events, no major turning points, no bold initiatives. There was little drama of any kind, but the lack of drama was more than offset by the period's undreamed-of prosperity.

Loose Threads

One figure dominated Wisconsin Electric during the postwar years: Gould W. Van Derzee. In March, 1945, months before the fall of Japan, the board had granted S.B. Way's request for "relief from the exacting duties of the presidency." (He was 70 at the time.) Way moved up to the newly created chairman's post, where he remained a definite influence, and Van Derzee moved into the president's office. As Way's chief assistant since 1913 and the system's general manager since 1934, WEPCO's "new" chief executive had little need for on-the-job training. Dignified, deliberate, and a good deal calmer than Way, this proper gentleman guided the company with a steady hand until his own retirement in 1962.

Van Derzee was soon alone at the top. On September 20, 1946, S.B. Way, his friend James Shaw (WEPCO's general counsel), and their wives were on a trip to the company's hydroelectric properties in Upper Michigan. A few miles north of Green Bay, their 1941 Cadillac collided head-on with a farm truck that had pulled out from the opposite lane to pass. Way, in the front passenger's seat, was killed instantly, and his companions were seriously injured. The accident was front-page

An engineer and a gentleman, Gould Van Derzee headed Wisconsin Electric during a period of unparalleled prosperity.

news in the state's newspapers. As the head of Wisconsin's largest utility for almost 20 years, Way's name was practically synonymous with electricity.

Gould Van Derzee would make his own name as a utility executive in the postwar period. Before he got down to the business of growth, however, Van Derzee and his colleagues had to tie off two loose threads that had been dangling since the 1930s. The first was Wisconsin Electric's separation from the North American Company. A foregone conclusion for years, the divorce was final in 1947, when North American retired some of its own common stock and gave shareholders an equivalent interest in its oldest subsidiary. WEPCO's ownership base soared instantly from 8,000 stockholders to 43,000. (Residents of New York, North American's home state, owned roughly one-fourth of the shares, followed by Wisconsinites with one-fifth.)

Already battling a steep decline in ridership and revenue, the transit system was buried by the blizzard of late January, 1947. (Paul Sass)

The transfer had absolutely no impact on the company's operations, but it did lead to sweeping changes on the board of directors. As North American appointees stepped down, the balance of power shifted from corporate insiders to outside directors who ran some of the largest businesses in the utility's service area. By 1948 six of WEPCO's nine directors were outsiders, and they included such local luminaries as William Coleman, chairman of Bucyrus-Erie; Harold S. Falk, head of the family-owned gear business; Charles Coughlin, president of Briggs & Stratton; meat-packer Michael Cudahy; and Albert Puelicher, president of M&I Bank. The predominance of outside directors gave Van Derzee both a fresh perspective on his own company and closer ties to Milwaukee's business community.

The second loose thread, and by far the more important, was the transit system. In May, 1947, Wisconsin Electric announced that the Transport Company, a wholly owned subsidiary since 1938, was for sale. The nominal reason for the action was the Holding Company Act of 1935. Without the umbrella provided by North American, WEPCO was a holding company in its own right, subject to the same law that was dismembering its former parent. Gould Van Derzee declared that the Transport Company was "not permanently retainable" under current rulings, and *Rail & Wire* lamented that the decision "could not be avoided and was taken with a great deal of reluctance."

The regrets were nothing more than crocodile tears. The federal government did not begin its review of Wisconsin Electric's operations until

1950, three years after the transit system was put on the block, and its primary focus then was the gas business, not transportation. The company was, in effect, surrendering before the first shots were fired. Wisconsin Electric had decided to leave the transit field in the mid-1930s, and it had taken a global war to postpone definitive action. The announcement of 1947 may have surprised the riding public but, from the company's viewpoint, it was long overdue.

The directors had scheduled a meeting for July 21, 1947, to open bids. To their profound disappointment, there were no bids to open. Without the artificial stimulus of war, the transit system had lapsed into its long-term pattern of failing ridership and falling revenue. The Transport Company's net income dipped from its wartime high of $863,000 in 1942 to a mere $287,000 five years later, and passenger traffic fell one-third between 1945 and 1952. No one, apparently, had both the nerve and the resources to take over a system that was in obvious decline. Until a buyer materialized, WEPCO had little choice but to keep managing a system it no longer wanted.

To their credit, company officials did not neglect their troubled stepchild. Public transit was still a public necessity, and the Transport Company labored to keep pace with Milwaukee's postwar expansion. The system had already been stripped of its interurban arms, but miles of new bus routes were added to serve the booming suburbs, and lines in the older sections of town were steadily modernized. The system's antiquated streetcars (the newest dated to 1929) had become an acute embarrassment, even to those inside

Streetcars came to be viewed as the dinosaurs of mass transit. The company used this 1945 view of "traffic tie ups" at 12th and Vliet in ads promoting its modernization program. (Milwaukee County Transit System)

Trackless trolleys were the replacements of choice after World War II, followed by gasoline and then diesel buses. This little Ford bus fed the Third Street trolley line. (Russell Schultz)

the company. In 1948, when trackless trolley service came to North Third Street, *Rail & Wire* praised the demise of the "outmoded" streetcars that had "caused passengers and motorists delays and annoyance." Roy Pinkley, head of the transit system since 1925, described the steel-wheeled car as an anachronism. "It has long been demonstrated," he wrote in 1952, "that street cars do not belong in modern traffic."

Pinkley tried to satisfy the public's clamor for "modern rubber-tired vehicles." Between 1946 and 1951, the number of cars in the fleet plummeted from 435 to 204, while the number of buses and trackless trolleys rose from 576 to 917. "Every substitution of rubber-tired vehicles," wrote Pinkley in 1951, "has been made at the request of civic associations and the Common Council." As the modernization effort gathered momentum, hundreds of vintage streetcars made their final runs to an old quarry east of Waukesha, where they were burned and then cut apart for scrap. Just imagining the holocaust is enough to give a railfan nightmares.

The modernization program kept the transit system in marketable condition, but years went by without even a nibble. In 1951 Wisconsin Electric offered its transit subsidiary to the City of Milwaukee. Municipal ownership had, after all, been a battle cry in earlier years, and the city's voters had elected a socialist mayor, Frank Zeidler, in 1948. Zeidler favored a metropolitan transit authority, but few aldermen shared his enthusiasm. Despite multiple fare hikes since 1947, the Transport Company's bottom line had shown little improvement. In 1952 a Common Council committee quietly turned down the offer:

> *It is the judgment of the committee that ownership and operation of the community's transit system by the present private company or an equally stable private company is undoubtedly preferable to any type of municipal operation.*

With streetcars disappearing fast, the old "trainmen's room" in the Fond du Lac Avenue car station was increasingly a gathering place for bus drivers. (Irwin Scroggins)

As the switch to rubber tires continued, hundreds of old streetcars were dispatched to a quarry near Waukesha and unceremoniously burned. (Milw. Co. Transit System)

Help was on the way. In October, 1952, following two years of sporadic negotiations, WEPCO agreed to sell the Transport Company to a new firm: the Milwaukee & Suburban Transport Corporation. M&ST was a four-man syndicate whose members included two Chicago lawyers, the Chicago manager of a gas and fuel company, and a Milwaukee coal dealer. The new owners called their acquisition an "outstanding" transit system serving "a great community," and they pledged to "continue and improve the present record of fine performance." For $10 million (with $4 million down), they took over the entire system, lock, stock, and management. The sale produced a book loss of $8 million for Wisconsin Electric, and a tax write-off of $3.5 million. The deal was closed on December 30, 1952, and WEPCO was finally out of the business that had been its reason for being in the days of John I. Beggs.

The subsequent history of the transit system has been generously chronicled elsewhere, but it clearly followed lines laid down in the early postwar years. Milwaukee & Suburban Transport continued to modernize its fleet, shifting steadily to the diesel buses that had first appeared in 1950. The last streetcar made its farewell run in 1958, and trackless trolleys vanished in 1965. (By the time Schuster's Christmas parade made its final appearance

WEPCO finally found a buyer for its troubled transit system in 1952. Representing the new owners, J. Roy Browning (seated second from left) shared a document with Roy Pinkley, who headed the system for one more year.

It was trucks, not trolleys, that provided the transportation for Schuster's Christmas parade in its waning years. Santa and company traveled down Center Street during their last ride in 1961. (Russell Schultz)

in 1961, Santa Claus and Billie the Brownie were transported by truck.) Although it was hardly a gold mine, the metropolitan bus system operated with relative success in a perilous period for mass transit. In 1975, finally, after a few years of crisis under private ownership, the system was purchased (with major federal assistance) by Milwaukee County. Public transit was officially public transit.

More than 40 years after Wisconsin Electric's departure from the transportation field, surprising traces of the original system remain. They linger in the routes and numbers of today's bus lines, in the old interurban roadbeds (now used as power corridors and bicycle paths), and in the miles of steel rail that surface periodically in street reconstructions. Perhaps the most telling sign of past excellence resides, ironically, in Milwaukee County's freeway system. Miles of old Rapid Transit lines lie buried beneath local expressways, whose planners knew a good route when they saw one.

"Live Better ... Electrically"

With the dissolution of the North American Company, Wisconsin Electric gained more control over its own destiny and a welcome respite from public controversy. With the sale of the Transport Company, the utility cured a chronic headache and freed its managers to concentrate on their central task: selling electricity. That task had never been easier. After 15 years of depression and war, consumers were starved for new homes, new cars, new appliances, and new creature comforts. The postwar economy was an explosion waiting to happen, and the decades after 1945 witnessed multiple booms: in housing, in population, in factory output, and in power consumption. Not since the 1920s, when the electrical base load was much smaller, had conditions been so ripe for growth. Mirroring national trends, the WEPCO system's annual output grew at an average rate of more than 7 percent in the two postwar decades. The utility's output doubled every 10 years, reaching 4.78 billion kilowatt-hours in 1955 and 9.71 billion in 1965. Growth on the bottom line was even more impressive. The system's net income finally passed the 1929 record ($7.58 million) in 1950 and continued climbing to $12.16 million in 1955 and $27.77 million in 1965 — an average growth rate of more than 9 percent annually for 20 years!

As in earlier periods, heavy industry led the way. War work had swollen the region's industrial capacity and, after a troublesome period of retooling, the factories of southeastern Wisconsin turned all of that capacity to peacetime production. Milwaukee, whose products ranged from motorcycles to machine tools, boasted the eighth-largest industrial output in the nation, and manufacturing engaged more than half the community's work force. To the south were giants like J.I. Case and Johnson Wax in Racine and American Motors in Kenosha, which grew out of the 1954 merger of Nash-Kelvinator and Hudson. All depended on Wisconsin Electric; the vast majority of manufacturers had scrapped their own power plants before or during the war. Large industries made up fewer than 1 percent of the utility's customers, but they consumed roughly 35 percent of WEPCO's power output in a typical year.

As the economy gathered steam, commercial customers enjoyed a boom of their own. New supermarkets installed new refrigerators and freezers, restaurants updated their kitchens with electric appliances, and WEPCO tutored contractors in "modern lighting fundamentals." Venerable shopping districts like North Third Street and Mitchell Street more than held their own after the war, but the general trend was away from older neighborhoods. With the opening of Southgate in 1951 and Capitol Court in 1956, the region entered the era of the modern shopping center, which offered plenty of free parking and a dazzling variety of stores. Both of the new centers consumed as much current as a small city's downtown.

Wisconsin Electric actively courted industrial and commercial customers after the war, but its major focus was the residential market, which offered both greater profits and greater potential for growth. After 15 years in a coma, the construction industry revived in the postwar period. The region's housing shortage reached critical proportions as veterans returned from war to start new families, and builders were hard-pressed to keep up with the demand for housing. In the 1950s, a particularly good decade for contractors, the number of housing units in the four-county Milwaukee area mushroomed from 298,835 to 400,684 — a 34-percent increase. Much of the growth took place in the City of Milwaukee, whose land area doubled during Frank Zeidler's tenure as mayor (1948-1960). But federal loan programs helped to steer home-builders to outlying areas, which quickly assumed new identities as suburbs. From Bayside to Brookfield and from Elm Grove to Oak Creek, a rash of incorporations changed the political map of southeastern Wisconsin after the war. With the incorporation of Greenfield in 1957, the "iron ring" of suburbs, as Frank Zeidler called it, finally closed around Milwaukee, and the wave of home construction continued to surge outward.

Wisconsin Electric had been planning for the housing boom since the war years. In the spring of 1944, sure that the Allies would prevail, WEPCO began to host meetings of the "Plan Your Home Club" in the Public Service Building auditorium. Speakers covered topics like site selection, design, furnishings and, of course, "the mechanical requirements for modern electric living." *Rail & Wire* explained the need for the club:

> *There are thousands of folks in the Milwaukee area who are eagerly awaiting the resumption of home building. But in their eagerness to get started, they are apt to rush into the project with the result that important home features, "big and little," may be overlooked.*

When the boom did come, all of the new homes needed new wiring, new lighting, and new appliances. Demand for electricity grew entirely on its own, but Wisconsin Electric boosted its load with an aggressive promotional effort. In 1945 Gould Van Derzee had looked ahead to emerging postwar technologies, including "the electric sink with garbage disposal unit and

The postwar boom created vast new markets for electricity, typified by the Bayshore Shopping Center (left) and sprawling subdivisions like the one at 76th Street and Hampton Avenue. (both Milwaukee Public Library)

dishwasher, the electric blanket, television, the precipitron to filter dust and pollens from the air used to heat your home, and the sterilamp installed in the heating system to kill all airborne germs and bacteria." Most of those futuristic devices did become fixtures in American homes, and dozens of others vied for attention: electric frying pans, can openers, charcoal lighters, shoe polishers, and lawn mowers. There were 166 electrical appliances on the market in 1963 (up from 13 in 1930), and they included such indispensables as the electric scissors, the face patter, and the cradle rocker.

Wisconsin Electric stuck to the basics, with particular emphasis on major appliances sold through independent dealers. Soon after the war, WEPCO offered $25 rebates to area dealers for every electric range or water heater they sold. Sales of both appliances doubled between 1947 and 1948, climbing to 6,000 ranges and 5,300 water heaters, and business continued at that level for years. Electric clothes dryers were the newest wrinkle in 1952; Wisconsin Electric provided free installation of the first 1,400 sold in its service area. In the late 1950s, the utility sponsored a continuing round of promotions — Carefree Cooking Carnivals, Dryer Delights, and Hot Water Jubilees — that offered money-back guarantees, free 220-amp circuits, or discounts on upgraded wiring for purchasers of major appliances. The wiring promotions were targeted to owners of older homes, most of which lacked adequate "housepower" for the full range of new appliances. Beginning in 1954, Wisconsin Electric sponsored a "Wire on Time" program that allowed homeowners to update their electrical service for as little as a dollar a month.

The goal of the various promotions was complete penetration of the residential market. In 1959 WEPCO launched its "Medallion Homes" program to recognize showcases of "electric living." Winners of the Gold Medallion had to meet the all-electric criterion, right down to their heating systems. The company competed head-to-head with coal, gas, and oil firms for residential heating customers, with mixed results. In 1965 only 638 homes in the region (including some without chimneys) were totally reliant on "flameless" electric heat.

Exhibitors at the Racine Home Show displayed the full range of home appliances available in 1947...

... while contractors tried to make sure that homeowners had "Adequate Wiring" to use them all. The wiring campaigns grew into the Medallion Homes program in 1959.

Whether they lived in Medallion Homes or coal-heated cottages, all residential consumers heard the same message: Live Better ... Electrically. That slogan became the industry's national theme in 1956, and Wisconsin Electric carried it forward on several fronts. Home Service Bureau specialists stepped up their schedule of classes and demonstrations, covering everything from bathroom lighting to holiday cooking. (One imaginative class was tailored to foreign-born "war brides" in Racine; the newcomers were tutored in "American eating habits and ... how to prepare wholesome and economical menus for their American husbands.") Consumers gladly accepted the company's help in living better. In 1954, a fairly typical year, Home Service Bureau employees handled 36,000 phone calls, served 125,000 drop-in visitors, made 6,000 house calls, drew plans for 1,500 kitchens (a free service), and put on demonstrations that attracted 40,000 people. The Christmas cooky display, featuring treats like Sherry Date Strips and Grandma's Fruitcake, remained a seasonal staple, and local homemakers vied to collect complete sets of WEPCO's Christmas Cooky Books. The number of books given away annually rose from 40,000 in 1952 to 100,000 in 1961.

The 1955 Christmas cooky display, like those before and after, attracted enthusiastic crowds ...

... and "Mary Modern" tutored television viewers in the art of electric living.

Wisconsin Electric used a variety of other means to keep its message before the public. Reddy Kilowatt continued his role as the company's ubiquitous "electrical servant." (Reddy earrings, oven mitts, lighters, tie clasps, and playing cards were available in the Public Service Building's retail store.) WEPCO also employed new media in its "mass selling" effort. *The Kilowatt Hour,* a mainstay of Milwaukee radio since 1929, tapered off to 15-minute segments of dinner music and finally left the airwaves in 1952. It was a victim of television. There were 80,000 sets in the Milwaukee area by 1950, and Wisconsin Electric could hardly ignore the medium. The company joined other utilities in sponsoring nationally televised programs like *The Silent Service* and *Hollywood Palace,* but it also produced shows for the local market. The most memorable was undoubtedly *Electrical Living — with Mary Modern,* which premiered in 1955. For five minutes each day, a spokesmodel offered viewers "a variety of entertaining ideas for making home life better ... electrically." "Mary" was a regular guest at company events, generally appearing with a tray of cookies in her hands.

Wisconsin Electric's central message of the postwar period — "Live Better Electrically" — was up in lights on the Public Service Building. Note the Greyhound Bus sign.

Television programs, consumer classes, and appliance promotions certainly helped to stimulate demand but, as always, the price of electricity was a deciding factor. New power plants and more efficient generation pushed prices steadily downward in the postwar years. The average rate for residential energy dropped from 2.69 cents per kilowatt-hour in 1946 to 2.45 cents in 1950 and 2.16 cents in 1965 — a 20-percent decrease in 20 years.

As rates fell, residential use of electricity skyrocketed. Fueled by the regional building boom, the WEPCO system's residential base grew nearly 60 percent between 1946 and 1965, rising from 361,142 customers to 563,631. But it was the increase in consumption per household that produced the most staggering figures. The average residential customer used 1,542 kilowatt-hours per year in 1946, 3,113 kilowatt-hours in 1955, and 5,115 in 1965. Energy consumption per household more than tripled in 20 years! It was clear, as the figures rolled in, that Wisconsin Electric's emphasis on the home market had been well-placed. At the end of World War II, large industrial and commercial customers were WEPCO's mainstays, contributing substantially more to the system's gross income than residential accounts. By 1965 the ratio had been reversed. Residential customers, who used less power but paid higher rates, brought in 57 percent more revenue than large businesses — a total of $62 million annually.

The sales figures were heartening, but just as important, from the company's point of view, was a pronounced shift in WEPCO's public image. In 1946 Wisconsin Electric began regular and systematic surveys of customer opinion. Seventy-seven percent of the 1948 respondents reported that the company had a "good" or "excellent" reputation in their neighborhoods. In 1959, 93 percent of those surveyed agreed that electricity was "the best bargain in the household budget." Even natural enemies found kind words to say. When Milwaukee observed National Electrical Week in 1958, socialist Frank Zeidler issued a laudatory statement:

> *The electrical industry, a vital industry in its own right, shares the great honor of making possible other great industries.... The telephone, the automobile, the public transport system, the hospital appliances, and all of the other indispensables to modern living are offshoots of the electrical industry.*

The shift in local opinion mirrored national trends. As the excesses of the holding-company era faded into the past, "business-managed utilities" (the preferred term in the 1950s) enjoyed a new reputation as public servants. In 1943 only 41 percent of Americans favored private ownership of electric utilities; by 1955 the proportion had climbed to 56 percent, and it continued to rise. Wisconsin Electric found its newly benign public image a welcome change. After decades as the heartless villain in a political morality play, the company gladly took its place at the forefront of the parade of progress.

Powering the Future

The Wisconsin Electric system faced only one major problem after World War II: keeping one step ahead of the soaring demand for its product. Electricity could not be back-ordered and then shipped when it was ready; demand had to be met instantly, and every new household appliance, every new industrial machine added to a load that constantly threatened to outstrip WEPCO's capacity. The problem was especially acute just after the war. In June, 1945, two months before Japan's surrender, company officials had broken ground for a third 80,000-kilowatt unit at Port Washington. They hoped to finish the plant in two years, but a wave of postwar strikes added 18 months to the construction schedule, and WEPCO was in trouble. As a power shortage loomed, appliance promotions were temporarily canceled, advertising was restricted to "informational and institutional" messages, and the relatively few factories that still maintained independent power plants were urged to keep them in good working order. For employees with long memories, the situation was reminiscent of the power crisis that followed World War I.

The Port plant's third unit was finally dedicated on October 5, 1948, averting a major crisis. As the company returned to "active promotion of new business," Gould Van Derzee resolved not to get caught again. Postwar planning studies indicated a need to double the system's capacity every 10 years, and the result was an absolute blizzard of new building projects. Work on the fourth and fifth units at Port Washington was already well under way by the time the third unit was completed. For the next 20 years, WEPCO's engineers were like jugglers trying to keep several balls in the air at the same time.

With the completion of Port Washington's fifth unit on December 18, 1950, the plant reached its final capacity of 400,000 kilowatts. Construction had taken more than 20 years and cost nearly $50 million, but it had

Pressed for capacity, WEPCO leveled another bluff (left) and finished the Port Washington Power Plant's third unit in 1948.

given the system a new workhorse; the Port plant constituted nearly half of WEPCO's total capacity. Although it had been dethroned as the world's most efficient power plant in 1949, after a reign of 13 years, Port Washington was still a model of operating economy. In 1936 the first unit had required a record low of 10,329 BTUs (the standard measure of heat) to produce one kilowatt-hour of electricity. With steady progress in boiler and turbine design, the fifth unit needed only 9,714 BTUs in 1951.

Even before the building crews left Port Washington, Wisconsin Electric was looking south for new plant sites. In February, 1948, the utility bought 170 acres on the Lake Michigan shore south of Kenosha. Soon, however, the company's real estate agents found an even more promising site: 368 acres of lakefront land in the unincorporated Town of Oak Creek, 12 miles south and east of downtown Milwaukee. WEPCO bought the parcel in the autumn of 1950 and promptly announced plans for a new power plant with a capacity of 500,000 kilowatts. It would, the engineers decided, employ the same unit design pioneered at Port Washington; the finished powerhouse would be four plants in one, each with a capacity of either 120,000 or 130,000 kilowatts.

Ground for the first Oak Creek unit was broken on May 4, 1951. The Korean War was on at the time; military contracts were boosting the demand for industrial power, and WEPCO declared that the plant "must be rushed to completion as quickly as possible." Rushing proved nearly impossible. Construction crews had to push a 100-foot bluff into Lake Michigan to provide a footing for the plant and a dock for visiting coal boats; they ultimately moved 400,000 cubic yards of earth. Progress was further delayed by strikes in the steel industry and the building trades. Demand for power, in the meantime, was going through the roof. In 1952 the system's peak load (930,000 kilowatts) exceeded its capacity (879,610 kilowatts) for the first and only time in the postwar years. All that saved the region from blackouts was Wisconsin Electric's interconnection with neighboring utilities. Oak Creek's first unit was finally completed on September 30, 1953, 10 months behind schedule, and power began to flow through lines strung along the old interurban rail corridor between Milwaukee and Racine.

Oak Creek was the focal point of Wisconsin Electric's building program for the next 13 years. As the company labored to keep pace with the demand for energy, work proceeded on two or even three units at once. The second installment of the Oak Creek plant was completed on November 1, 1954; the third on December 4, 1955; and the fourth and final unit on October 22,

The spotlight shifted to Oak Creek in 1951. A massive landfill project (left) created a footing for the north plant, whose first unit opened in 1953.

As the demand for power continued to soar, WEPCO began work on a second Oak Creek plant just south of the first. By the time of its completion in 1967, the entire complex (right) had a capacity of 1,670 megawatts — more than four times the output of Port Washington at its peak.

1957. With a capacity of 500,000 kilowatts, the finished powerhouse displaced Port Washington as the system's workhorse, and the Lakeside plant, a technological marvel in 1920, became a stand-by facility. In earlier years, the completion of a new unit would have been celebrated with long speeches, brass bands, and ceremonious ribbon-cuttings. There was no fanfare in the Fifties; WEPCO simply put one unit into service and moved on to the next. Addressing the company's veterans in 1956, Gould Van Derzee said, "We are now so used to the thought that we are to double our electrical load [every] ten years that everyone seems to take it for granted."

Absolutely no one questioned the need for more power in the 1950s. State authorities, whose historic relationship with utilities bordered on the adversarial, seemed to bend over backward to make sure that everyone had enough energy. The Public Service Commission routinely approved applications for new power plants within weeks of their submission. In 1947 Wisconsin's legislators outlawed utility strikes (an action overturned by the U.S. Supreme Court four years later), and in 1955 they made it easier for utilities to condemn land and protect long-term easements. Even financing was relatively easy. Wisconsin Electric funded much of its building program through sales of common stock, which was issued, on average, every two years. Utility stocks had become an investment of choice during World War II, and WEPCO's offerings were generally oversubscribed. The utility rewarded the faith of its investors; the price of Wisconsin Electric's common stock climbed from $11.75 per share in 1944 to more than $35 in 1958.

It had never been easier to build power plants, and WEPCO never slackened its breakneck construction pace. In June, 1956, Gould Van Derzee announced plans for a 1-million-kilowatt powerhouse just south of the original Oak Creek facility. Like its smaller predecessor, the plant was designed to house four units, each built around a 250,000-kilowatt turbo-generator supplied by Allis-Chalmers. Van Derzee called the project a demonstration of "our faith in the future of Wisconsin and the communities within our operating area." Work on the plant consumed the next 11 years, with the usual interruptions and design changes. The first two generators were upgraded to 275,000 kilowatts before installation, and the second pair (manufactured by General Electric) were rated at 310,000 kilowatts each. The south plant's four units went on line in 1959, 1961, 1965, and 1967. Operating in tandem with the four units in the north plant, they gave Oak Creek a final capacity of 1,670,000 kilowatts — or 1,670 megawatts. In the mid-1960s, as the number of zeroes reached unwieldy proportions, the megawatt (1,000 kilowatts) became the industry's preferred unit of measurement.

Power plant technology, of course, changed constantly in the postwar period. The general trend was toward higher steam temperatures and higher turbine speeds, both of which were made possible, in part, by advances in metallurgy. Controls underwent a comparable evolution; Oak Creek's seventh and eighth units featured solid-state boiler combustion and computerized control centers. Every unit was more sophisticated than the last, and the result was continuous improvement in operating economy. Despite the presence of some antiquated plants in the WEPCO system, the average amount of heat required to produce one kilowatt-hour of electricity dropped from 13,470 BTUs in 1947 to 11,079 in 1955 and 9,840 in 1965 — a 27-percent improvement in 18 years. Construction expenses forced the company to raise its rates slightly in 1953 and 1958 (the first rate increases since 1920), but improvements in generating efficiency were more than enough to sustain the long-term trend toward lower energy prices.

Technical advances transformed Wisconsin Electric's power side after World War II, but the most obvious change was a dramatic increase in the scale of generation. WEPCO moved from 80-megawatt generators in 1950 to 310-megawatt units in 1965; Oak Creek's eighth unit could turn out as much electricity as the entire Lakeside plant in its prime. As the power plants at the heart of the system produced more energy, there was corresponding growth in the system's arteries — its transmission lines. WEPCO's standard line voltage jumped from 132,000 volts in the 1950s to 230,000 in the 1960s. Higher line voltages, in turn, made it easier to transport power over long distances. Plants in the Milwaukee area served customers as far away as Upper Michigan, and there was a growing trend, encouraged by federal defense planners, toward interconnection with other utilities.

Erecting crews hoisted one of the frames for a 345,000-volt transmission line near Appleton in the mid-1960s. The line linked Wisconsin Electric with other utilities in a new regional power pool.

WEPCO had been sharing lines with its Wisconsin neighbors since the early 1900s, but interconnection entered a new dimension after the war. The United States developed a truly national power grid, and Wisconsin Electric was a full participant. In 1952 the utility re-established its tie with Commonwealth Edison at the Illinois border. In 1963, after prolonged negotiations, WEPCO began work on its portion of a 345,000-volt line running from Chicago to St. Paul through Milwaukee and Appleton. One year later, the company became a charter member of the Mid-America Interpool Network (MAIN), which combined the power resources of nearly all the utilities in a 10-state region. The result was superpower revisited. Interconnection recalled the holding-company empires of the 1920s, which brought entire regions together under one corporate umbrella. But MAIN was an alliance rather than an empire; each member bought and sold energy according to its needs. The advantages to the utilities (and their customers) were obvious: Interconnection provided backup power in times of emergency, a profitable outlet for reserve capacity, and the deferral of millions of dollars in construction costs.

The physical improvements of the postwar years, from regional high-voltage lines to state-of-the-art power plants, were anything but cheap. Wisconsin Electric's capital spending averaged $40 million per year in the decade after 1954, more than $200 million in 1996 dollars. But WEPCO got what it paid for; the system's total capacity soared from 571 megawatts in 1946 to 1,253 in 1955 and 2,140 in 1965 — a 275-percent increase that stayed just ahead of the 250-percent rise in peak demand. The result was nothing less than the economic well-being of the system's entire service area. Building for a future that always arrived sooner than expected, Wisconsin Electric made sure that a booming region had all the power it needed to grow.

"Wisconsin Natural" and Wisconsin Michigan Power

The power plants at Oak Creek, Lakeside, and Port Washington served the entire WEPCO system, from the Illinois border to Upper Michigan, but the system included two subsidiaries that maintained strong geographic identities of their own: Wisconsin Gas & Electric and Wisconsin Michigan Power. Both companies kept pace with their parent's growth in the postwar years, amplifying a boom that reached every corner of the service area.

WEPCO's largest subsidiary was radically restructured early in the period. Since its formation in 1912, Wisconsin Gas & Electric had been a

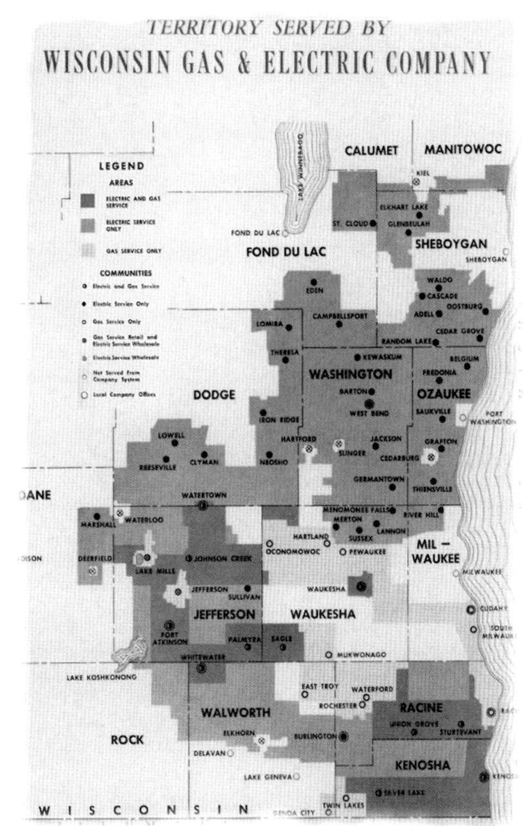

Wisconsin Gas & Electric's 1948 service area was, to say the least, complex. Merging its electrical properties into WEPCO simplified matters.

combined utility, serving both gas and electric customers in much of southeastern Wisconsin. On June 9, 1950, Wisconsin Electric took over WG&E's entire electrical side, which provided service to thousands of customers in a wide ring around Milwaukee. The move was made, WEPCO officials explained, in the interests of "more efficient and economic operation of the consolidated system," but there were other motives. With the dissolution of the North American Company, issues of integration were receiving more attention. Merging WG&E's electrical properties into WEPCO's strengthened the system's status as a "single integrated utility." Federal investigators found no reason to tamper with the new structure; in 1955 Wisconsin Electric won its first exemption from the Holding Company Act.

Federal mandates were an important stimulus, but the most potent reason for WG&E's break-up was the advent of natural gas. The gashouse on Racine's lakefront had been the company's most important asset in the early years, and manufactured gas had remained, like the steam heating system in downtown Milwaukee, a relatively small but profitable adjunct to the electrical business. Natural gas changed all that. Rumors of a gas pipeline from the Southwest had been circulating since the early 1930s, when a Chicago firm proposed a northerly extension of its own service. (S.B. Way testified against the extension in 1931, arguing that natural gas was a bad bargain for consumers and a threat to WG&E's gashouse employment.) The Depression and World War II delayed action of any kind until 1946, when William Woolfolk, a Detroit utility magnate, formed the Michigan-Wisconsin Pipe Line Company. Woolfolk unveiled plans to build a new pipeline from the Hugoton gas fields in Texas, Kansas, and Oklahoma to existing utilities in Michigan and Wisconsin. The utilities would act as distributors, buying wholesale gas for resale to customers at retail rates. The pipeline quickly passed muster with federal officials, and work began at the Texas-Oklahoma border in December, 1947.

Natural gas was clearly the fuel of the future. It had twice the heating value of gas extracted from coal and, with the opening of new reserves, it was significantly cheaper as well. In 1947, months before work on the pipeline started, WG&E received permission from state regulators to convert to natural gas. Two years later, as construction crews inched their way toward "Wisconsin Junction" at Joliet, the utility signed a 25-year supply contract with Michigan-Wisconsin Pipe Line.

The company was hard-pressed, in the meantime, to keep up with the demand for manufactured gas. The aging gashouse in Racine was already stretched to capacity, and modernization there had all the appeal of an investment in buggy whips. WG&E was forced to turn down every request for new service (including one from the Nash-Kelvinator factory in Kenosha) and to build a number of small propane plants to meet peak demand. The long wait finally ended in November, 1949, when the 1,500-mile pipeline reached WG&E territory in Waukesha. (Every mile of steel pipe was supplied by Milwaukee's own A.O. Smith Company.) Conversion of the entire WG&E system was completed in February, 1950.

With the addition of a hot new product and the subtraction of its electric properties, Wisconsin Gas & Electric needed a new name. On May 17, 1950, a few weeks before the transfer of its electric side, WG&E was rechristened the Wisconsin Natural Gas Company. "Wisconsin Natural" came into existence with 57,000 gas customers in a broad corridor stretching from Racine and Kenosha northwest to Watertown and Whitewater. Natural gas quickly attracted thousands more. New industrial accounts multiplied (among them the massive Ladish forge shop in Cudahy), and gas became a fuel of choice for residential stoves, water heaters, and furnaces. In its first 10 years, the Wisconsin Natural system grew from 57,000 customers to more than 76,000, from 641 miles of gas mains to 1,266, and from $4 million in revenue to $13 million. The growth in total consumption was little short of astounding. Between 1950 and 1965, Wisconsin Natural's customer base increased 75 percent, but its gas sales, measured in therms, jumped 1,500 percent! The result was a larger role for the subsidiary in the success of its parent. The gas business accounted for just 6 percent of the WEPCO system's

The coming of the natural gas pipeline transformed the old WG&E system. The company was rechristened "Wisconsin Natural" in 1950, and "Steddy Flame" (inset) became Reddy Kilowatt's counterpart on the gas side.

the Environmental Protection Agency to try another technology. A small-scale chemical plant was installed at the Valley powerhouse to capture the sulfur dioxide from one boiler and convert it to a usable chemical. The test was a technical triumph but a financial flop; the removal facility required too much space, too much power, and too much money to merit commercial application. The work has since continued in other directions, but a final solution to the SO_2 problem remains elusive.

Although air pollution was the major environmental issue confronting the company, there was no shortage of others. In the mid-1960s, when Lady Bird Johnson persuaded her husband Lyndon to support "the urgent task of conserving America the Beautiful," the public took a new interest in aesthetics. Billboards, power lines, parking lots, and commercial strips were lumped together as "visual pollution," and Wisconsin Electric tried vigorously to soften its own impact on the landscape. Declaring itself a champion of "Beautility," the company put up "slenderized" transmission poles and "sky gray" overhead equipment. Underground residential wiring, which debuted in a New Berlin subdivision in 1958, became the system's standard for new construction in the late 1960s. Although they had a functional beauty all their own, new electrical substations, none of them staffed, were hidden behind facades that resembled ranch houses or storefronts, right down to the curtains in their windows. In 1967, showing an early sensitivity to mounting criticism, the company promoted its "new look" as an answer to "the cynics of the wayout generation."

The "Beautility" policy was also applied to natural resources under the system's direct control. Wisconsin Michigan Power had begun to plant tree

"Beautility" became a priority in the 1960s. Substations like this one in Elm Grove were built with residential facades ...

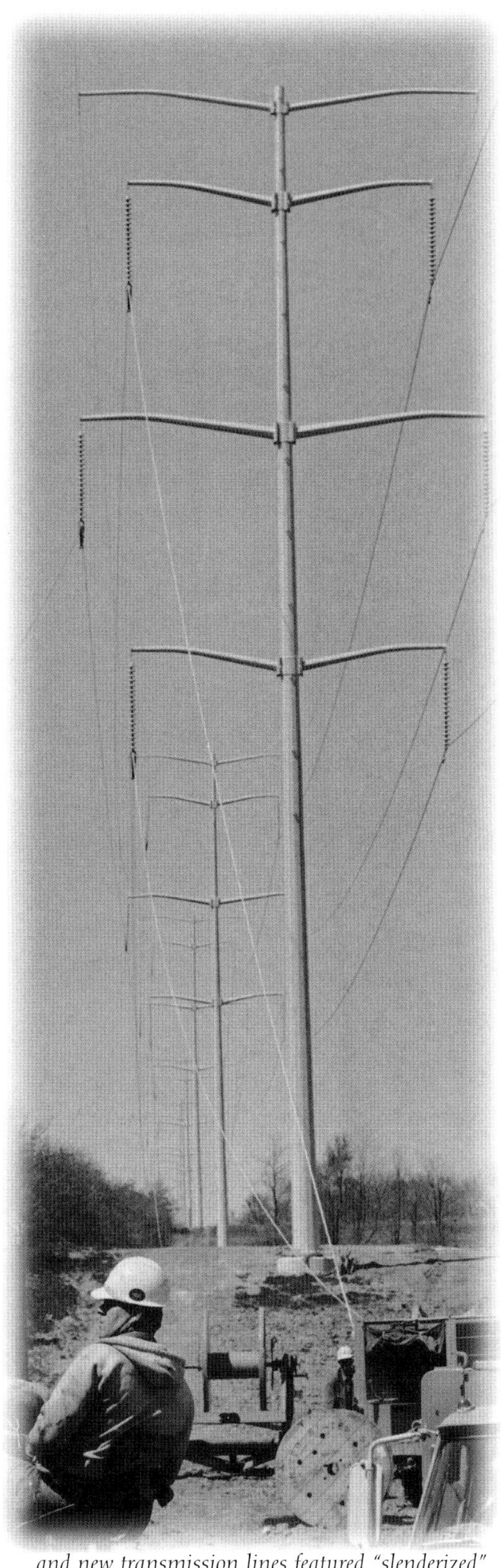

... and new transmission lines featured "slenderized" poles and "sky gray" paint schemes.

Beginning with the first seedlings in 1942, tree plantations in Wisconsin Michigan Power territory grew to cover thousands of acres.

In 1969 Gov. Warren Knowles (left) and John Quale signed an agreement for the preservation and public use of company lands in northern Wisconsin.

seedlings (principally red pines) on a few of its 42,000 acres in 1942; by 1970 the plantation had grown to 500,000 trees. In 1966 the subsidiary signed an agreement with the State of Michigan to develop limited recreational facilities (campsites, boat launches, and canoe portages) on its lands along the Menominee River; a similar compact with the State of Wisconsin followed in 1969. Closer to the system's population centers, WEPCO opened bicycle trails on some of the Milwaukee area's old interurban rail lines in the 1970s.

Opening its lands to public use helped to disarm some critics, but others chided the company for its treatment of another resource: water. Every steam plant uses copious amounts of cooling water in its generation cycle, and every steam plant returns that water to its source 10 to 20 degrees warmer. Thermal pollution became a hot issue in the late 1960s. Concerned about the possible effects of waste heat on aquatic life, federal officials proposed a ban on all "significant" discharges into Lake Michigan in 1970. One year later, Wisconsin adopted some of the most stringent thermal-discharge rules in the nation. Blasted by company lawyers as "arbitrary, capricious, and unenforceable," the rules might have forced WEPCO to equip all its steam plants with expensive cooling towers. (Resembling gigantic concrete snail shells, the towers circulate water at atmospheric temperatures before returning it to its source.) Wisconsin Electric consistently argued that its plants had no more impact on the lake than a day of sunshine, and Sol Burstein charged that public concerns amounted to "hysteria." In 1973, after two years of study, Wisconsin adopted regulations that its largest utility found somewhat less odious. Cooling towers, however, were included in plans for every subsequent power plant.

"No Nukes"

On issues ranging from fly ash to SO_2 and from visual pollution to thermal pollution, WEPCO was under constant fire during John Quale's presidency, but his most significant battles involved nuclear power. Few environmentalists showed as much passion as the "No Nukes" activists, and fewer still demonstrated such power to disrupt the utility's plans. The fact that nuclear energy was born under a mushroom cloud helps to explain the virulence of the opposition. Although turning an instrument of annihilation into a source of benevolent power had widespread appeal, nuclear fission never shed its dark side. The invisible, insidious effects of radiation, reaching down through generations, were the stuff of nightmares, and the sheer staying power of nuclear waste, which remains "hot" for 500 to 1,000 years, made the problem seem practically immortal.

Wisconsin Electric contended that atomic energy, under the careful control of the nation's utilities, posed "no serious threat" to anyone or anything. Sol Burstein once compared redundant nuclear safeguards to a man wearing a belt, adding a pair of suspenders, and then sewing his shirt to his pants. Such reassurances failed to sway the anti-nuclear forces, and a pronounced shift in the rule of law gave their viewpoint unusual leverage. In earlier years, anyone who objected to a utility's plans had to prove that those plans were not in the public interest. After 1970, with the advent of environmental impact statements and changes in case law, the burden of proof shifted to the utilities; they had to demonstrate that their plans *were* in the public interest. Environmental groups may have been perennially underfunded and understaffed, but they had the legal power to intervene in almost any case involving a utility.

Point Beach nuclear technicians began to load fuel in the plant's reactor on October 6, 1970, only hours after the AEC issued an operating license.

Opposition to nuclear power built slowly. In August, 1970, as the first Point Beach unit neared completion, a coalition of 34 Wisconsin environmental groups fired the first shot by petitioning to intervene in the plant's license proceedings. Their major concern was radioactive discharges, and the groups withdrew when WEPCO agreed to keep emissions well below existing federal standards. The Atomic Energy Commission issued an operating license for Point Beach on October 5, 1970, and the nation's nineteenth nuclear plant went into full-time operation on December 21 — four years and one month after groundbreaking.

Wisconsin Electric hailed the plant as "a culmination of historic progress" and a breakthrough equal to the completion of Lakeside 50 years before. "It opens the way," the company promised, "to higher levels of service reliability, conservation of fuel resources, economy in overall fuel expense and maximum compatibility with the environment." Point Beach was viewed as a panacea. After years of uncertain coal supplies and a shorter period of public outcry over pollution, WEPCO expected the plant to set new standards of efficient, economical and, above all, clean operation.

The second unit at Point Beach did not join the system so easily. On April 5, 1971, only a few months before the plant was finished, three groups — the Sierra Club, Businessmen for the Public Interest, and Two Creeks-based POWER (Protect Our Wisconsin Environmental Resources) — petitioned to intervene in the case. They charged that the company lacked both a workable emergency plan and adequate radiation controls, but their principal concern was thermal pollution. New court rulings mandated a full review of their objections, and WEPCO was stymied. The AEC permitted the unit to operate at 1-percent power in May, 1971, but a commercial license awaited settlement of what promised to be a lengthy legal proceeding.

Demand for electricity, in the meantime, was threatening the system's capacity. On July 21, 1972, a heat wave forced the company to reduce its voltage by 2.5 percent and to ask local industries to turn off some of their machines. Citing the "procedural morass" that had idled a state-of-the-art nuclear plant, John Quale laid the problem squarely at the feet of the intervenors. On July 28, seven days after the record peak, the AEC permitted Point Beach 2 to raise its output to 20 percent of full power. Hearings in the case continued all the while, and the final transcript, sprawling over 6,000 pages, made a Russian novel seem like a business memo. It was not until March 8, 1973, after nearly two years of legal wrangling, that the intervenors' objections were dismissed and Point Beach was allowed to operate at full power. "Incredible," said John Quale. The new plant soon rewarded the faith that Alfred Gruhl and Gould Van Derzee had placed in nuclear power. It became, with Oak Creek, a mainstay of the system and one of the safest, most reliable nuclear plants in the world.

Despite the delays involving the

Legal challenges limited use of its second unit for two years, but the Point Beach Nuclear Plant was operating at full power by early 1973. It quickly joined Oak Creek as a mainstay of the WEPCO system.

second unit, Point Beach got in just under the wire. The rules governing atomic power were tightening like a noose, and public opposition was mounting by the month. Wisconsin Electric soon found the road to a nuclear future disappearing before its eyes. The timing was, from the company's point of view, unfortunate. In the early 1970s, John Quale and his executives still believed that demand would double every 10 years, and they looked ahead to a serious power shortage. Gas-burning combustion generators, whose turbines were driven by exhaust gases rather than steam, were already available to help with seasonal peaks; combustion units were installed at Oak Creek and Lakeside in 1968, and at Port Washington and Point Beach in 1969. But Wisconsin Electric needed a much larger plant to handle its future base load. On May 10, 1972, the company announced plans for a powerhouse with a capacity of 1,800 megawatts, twice the maximum output of Point Beach. A new era of cooperation was beginning; WEPCO entered the project in partnership with three neighboring utilities: Wisconsin Power & Light, Wisconsin Public Service, and Madison Gas & Electric. (Those neighbors already shared an interest in the Kewaunee Nuclear Plant, which was taking shape a few miles north of Point Beach.) Wisconsin Electric was the lead partner, holding a 62.5-percent stake in the project and acting as the group's agent.

Groundwork for the joint venture was laid with the utmost care. It was not until the summer of 1973, a year after the announcement, that the partners made a commitment to nuclear fuel, and it was not until the summer of 1974 that they selected a site. (A state-imposed moratorium on Lake Michigan power plants had forced the search inland.) On June 25, 1974, John Quale announced the winner: a 1,410-acre parcel on Lake Koshkonong, a flowage of the Rock River near Fort Atkinson. Anticipating a barrage of criticism from environmentalists, Quale declared that the plant was absolutely necessary:

Encouraged by the success of Point Beach, Wisconsin Electric joined neighboring utilities in planning for an 1,800-megawatt nuclear plant near Lake Koshkonong in southern Wisconsin.

I'd rather face a handful of critics now than try to explain to three million people why electricity had to be rationed in 1981.... There's no time for a breakthrough in fusion. There's no time for a breakthrough in solar power. There's no time for a breakthrough in energy storage. There's no time for population control. The population is here. The population is your children and my children who are already in school, and they will need electrical power.... Nuclear power is more than an acceptable power source — it is a preferable source.

Quale expected criticism, and he got it. The Koshkonong plant became the eye of a rhetorical hurricane, with charges and countercharges swirling in all directions. Anti-nuclear activists raised the specter of meltdowns and runaway radiation. Madison became the first city in the nation to go on record against atomic power in any form. Environmentalists declared that the plant would cause irreversible damage to wetlands, wildlife, and the lake itself. (Koshkonong's average depth was a little over five feet.) The utility partners counterattacked with a caravan of public speakers, a door-to-door canvass of the plant's potential neighbors, and an information center in Fort Atkinson. Voters in the Town of Koshkonong first endorsed a ban on nuclear plants and then reversed themselves. State regulators added their voices to the debate. In late 1974 the Public Service Commission barred the utilities from placing firm orders for equipment and warned that money already spent on the project (nearly $14 million) was at risk. Work on the environmental impact statement continued in the meantime, with no final rulings expected for at least two years. Wisconsin Electric stuck to its guns through John Quale's presidency but, at the end of 1975, the prognosis for Koshkonong was guarded at best.

Plans for a more conventional plant encountered a somewhat less chilly reception. Koshkonong was planned as a base-load facility, with completion of its first unit scheduled for 1981, but WEPCO's demand studies indicated the need for an "intermediate-load" powerhouse at least two years earlier. The company briefly considered placing the plant in the Town of Paris, west of Kenosha, but officials quietly scrapped that plan when 84 percent of the town's voters signed a petition against it. On June 19, 1974, after further exploration, WEPCO unveiled plans for a 1,160-megawatt coal plant in the Town of Pleasant Prairie, a few miles southwest of Kenosha. John Quale sounded a familiar theme in making the announcement: "We would rather face criticism today for building the necessary generating plants than know in our conscience that we would be failing ourselves and our children in the years ahead."

Pleasant Prairie officials welcomed

In 1974 John Quale announced plans for an intermediate-load plant in Kenosha County. Its site (right) was a sprawling tract of farmland in the Town of Pleasant Prairie.

the project, but environmental groups and state officials expressed a skepticism that was becoming routine. Bowing to concerns about thermal pollution, WEPCO budgeted $42 million for cooling towers attached to the plant. Another $1 million was spent on a voluminous environmental impact statement. Groundbreaking was further delayed when the Department of Natural Resources questioned the plant's potential impact on air quality. The case ended up in court, and Pleasant Prairie was still unfinished business at the end of John Quale's presidency.

The controversies surrounding Point Beach, Koshkonong and, to a lesser extent, Pleasant Prairie illustrated one of the key characteristics of the environmental debate: its almost-complete lack of a middle ground. Like so many other "debates" of the late 1960s and early 1970s, the discussion was polarized and personalized, with a distinct us-vs.-them mentality emerging on both sides. The primary differences between Wisconsin Electric and its critics were, at root, ideological. Filled with quasi-religious fervor, the most ardent environmentalists openly opposed economic growth of any kind. What these latter-day druids wanted was nothing less than a return to a pre-industrial Eden, the untrammeled garden supposedly enjoyed by our ancestors. Wisconsin Electric, in their view, was a malevolent plunderer, both symbol and symptom of a power-crazed society careening toward its own destruction.

Company officials gladly returned the fire. John Quale was hostile, even truculent, in dealing with environmentalists, blasting the "pseudo-science"

Resistance to new power plants, particularly those using nuclear fuel, gathered force through the 1970s.
(Milwaukee Public Library)

and "distortions" of "fringe groups" who wanted to stop his system in its tracks. "Criticism requires only a tongue," he said during the Koshkonong debate, "but no discipline, and very little knowledge." Sol Burstein, characteristically, went even farther, charging that new federal regulations amounted to "ecological embezzlement." The utility's stand was based on an ideology of progress; WEPCO's leaders firmly believed that abundant electricity was the power behind America's prosperity. They could hardly comprehend, much less accept, the no-growth philosophy of the environmentalists, a point that John Quale made clear at the 1970 annual meeting:

> *There are some people who suggest that the one way to correct the environmental situation is to stop abruptly any growth in the electric utility industry — to stop us from promoting the use of electric service. This, I submit, is not possible nor practical. It flies in the face of reality. Since the beginning of man, progress in our society and in our world, our social and economic well being, have required constant and progressive growth and improvement. Without it we would be a static and, eventually, a decaying society.*

The ideological battle lines between WEPCO and its critics were clearly drawn, but the signals from the utility's own customers were profoundly mixed. People obviously wanted power, a desire they made clear in their ever-increasing demand for electricity to run their televisions, dishwashers, air conditioners, garage door openers, washing machines, electric blankets, and stereo systems. But there was genuine concern about the destructive side effects of progress, and that concern found expression in an imposing body of environmental laws. The American people, acting through their legislators, sought to rein in the very industries that had provided them with such an abundance of comforts and conveniences. The body politic resembled, at times, an alcoholic trying to curb the output of a distillery.

John Quale, for his part, found nothing embarrassing about the nation's consumption of energy. "There is no doubt," he said in 1974, "that we have been profligate users of energy for the last several decades. This is not necessarily all bad, since it is probably the single greatest reason why our standard of living is the highest in the world." Quale did acknowledge that the United States had "used up and, yes, sometimes wasted valuable resources," but he pleaded for an understanding of his company's role in society. "The public must recognize," he insisted, "that environmental programs must be fairly balanced with our prime obligation of furnishing reliable and reasonably priced electric service."

Balance of any kind was a scarce commodity during Quale's years at the helm. The irresistible force of Wisconsin Electric's expansion met the immovable object of environmental protest, and the result was constant friction. In a long view, the controversies marked a return to one of the system's oldest traditions. The utility had spent most of its history on the defensive, constantly deflecting charges that it was a heartless monopoly bent on fleecing the public. In comparison with the inflammatory rhetoric of socialist reformers or the violence of the 1934 strike, the protests of the 1960s and '70s were, if anything, relatively mild. After 20 years of postwar calm, they may have been shocking, but they were hardly unprecedented.

Although no one liked wearing a black hat, WEPCO's leaders tried to respond and adjust, however grudgingly, to new public expectations. Between 1963 and 1973, Wisconsin Electric spent nearly $30 million on air pollution control alone. In 1973 the company established a full-fledged Environmental Department, staffing it with limnologists, zoologists, meteorologists, and other specialists whose presence would have seemed bizarre a few years earlier. Seeking permanent answers to its problems, WEPCO funded research into nuclear fusion, coal gasification, cold storage, breeder reactors, and a host of other alternative technologies. But the single most important change was a shift in attitude. In 1971 John Quale noted that the utility industry's overriding goal had always been "service with the least cost." "We are now in a period of transition," he continued, "and are entering a new era, where the guidelines will become reliable service with reasonable cost and minimum environmental impact." During Quale's presidency, the environment was a factor in every decision the utility made, and environmental issues gained a primacy they have never lost.

The Economic Environment

The environmental movement was basically a revolution in attitudes. Americans developed a new sense of stewardship for the world around them, and American utilities changed accordingly. But Wisconsin Electric and its counterparts were changed even more profoundly by a revolution, or devolution, in the nation's economy. By the late 1960s, the protracted postwar boom had ended. Fueled by federal spending for both the Great Society and the Vietnam War, the economy overheated, producing a period of rising inflation, sharp recessions, and general volatility. For America's utility industry, the consequences bordered on the catastrophic. Environmentalists rocked the boat; economic turmoil nearly swamped it.

Wisconsin Electric had been riding an upward spiral since the end of World War II. More people used more power, leading WEPCO to build larger and more efficient plants. Larger plants meant lower unit costs, and so more people used more power. The cycle was practically self-generating, but its continued rise depended on a fragile set of conditions: stable construction costs, stable interest rates, stable fuel prices, and increasing productivity. The stability underlying the growth of America's utilities began to fray during Alfred Gruhl's tenure, and it unraveled completely under John Quale.

The utility business is capital-intensive by nature; gigantic power plants and sprawling distribution networks represent an enormous investment. Wisconsin Electric's assets passed the $1 billion mark in 1969, and the system's three newest power plants — Oak Creek, Valley, and Point Beach — accounted for more than a third of the total. The growing need for pollution-control equipment, coupled with soaring labor and material costs, made further additions exorbitantly expensive. The Oak Creek plant was built for roughly $121 per kilowatt, the Valley powerhouse for $143, and Point Beach (one of the nuclear industry's all-time bargains) for $176. Pleasant Prairie, by contrast, was budgeted at $345 per kilowatt, and the Koshkonong Nuclear Plant at $555. Both estimates, it turned out, were decidedly conservative.

WEPCO's construction program was creating pressures even before prices got out of hand. Between 1967 and 1971, the company's capital spending averaged well over $100 million a year, much of it borrowed in a period of tight money and rising rates. Wisconsin Electric's long-term debt more than doubled between 1965 and 1970, swelling from $230 million to $514 million, and the "imbedded cost" of that debt climbed from 3.63 to 5.68 percent. Interest payments rose alarmingly, from $8 million in 1965 to $31 million in 1970. With millions of non-productive dollars added to its expense base, the company's capital structure showed clear signs of strain. Wisconsin Electric's debt ratio reached a postwar high of 59 percent in 1969, well above the 50-percent mark considered prudent.

Nor could WEPCO count on improved efficiency to recoup the money it was spending, particularly on coal plants. Advances in boiler and turbine technology reached a plateau in the 1960s; the heat required to produce one kilowatt-hour of electricity bottomed out at 9,840 BTUs in 1965 and then began a gradual rise. There were also problems with the coal supply. The heating value of the "prehistoric sunshine" available to WEPCO was increasingly erratic, and even inferior

The scale of generation increased in the 1960s, as evidenced by the spindle of this Allis-Chalmers turbine being readied for shipment to Oak Creek. Construction costs rose even faster and operating efficiencies reached a plateau, putting severe economic pressure on the company.

coal was more expensive. The system's buyers paid an average of $6.97 per ton in 1965 and $8.72 in 1970, an increase of 25 percent in five years.

Other costs rose even more dramatically. Wisconsin Electric's state and local taxes climbed from $14 million in 1965 to $28 million in 1970, outpacing revenue growth by a wide margin. After leveling off in the decade following 1955, the company's work force began to grow again, increasing from 4,966 employees in 1966 to 5,631 in 1970, and payroll expenses grew accordingly. Both Alfred Gruhl and John Quale tried to hold the line on raises, which drew a sharp reaction from the utility's union locals. Power plant workers staged a 12-day strike during the Christmas season of 1966, forcing hundreds of management employees to spend their holidays in the plants, and the company's electrical workers walked out for 16 days in 1969. The settlements in both cases added materially to WEPCO's labor costs.

Whether it was dealing with contractors, fuel suppliers, lenders, tax authorities, or its own employees, Wisconsin Electric faced rapidly escalating costs in practically every aspect of its business. The cumulative result, barely perceptible at first, was a complete reversal of the rules that had guided the utility since 1945. Bigger was no longer better. Economies of scale had reached their tipping point, and unit costs were going up instead of down. Economic stagnation was the clearest sign of impending trouble. Although demand was still growing, the company's earnings were stuck at about two dollars per share between 1965 and 1971, a performance that John Quale criticized as "inadequate and relatively flat."

The inevitable result of weak earnings was a round of rate increases. In 1968 WEPCO won approval for its first rate hike since 1958, a 5.7-percent jump, and requests were filed practically every year thereafter. In 1971 the company issued an ominous warning: "Sharp climbs in virtually every cost of doing business, including the nonproductive but necessary costs of environmental protection, may signal the end of historic per unit cost reduction in the price of electric power."

Multiple rate increases led to more troubles. The consumer movement, tapping the same vein of energy that animated environmentalists and antiwar activists, was in full swing by 1970, and Wisconsin Electric became one of its favorite targets. Rate hearings, formerly the exclusive province of lawyers and engineers, attracted scores of protesters who castigated the utility as "greedy," "cruel," and "heartless." Russell Britt, the company's chief financial officer, was once treated to a "Consumer Theater" performance that ended with the cast throwing a pile of Reddy Kilowatt pins on his desk. Britt also heard a group of African Americans complain that virtually all blacks were required to make security deposits

Striking workers blocked access to the Lakeside Power Plant in 1966. Settling the dispute added one more layer to the cost pressures facing Wisconsin Electric.

The energy crisis of 1973-74 led to short-term gasoline rationing and long-term worries about America's economic stability. (Milwaukee Journal Sentinel)

before receiving service. (WEPCO responded; deposits were no longer required from anyone after 1973.) Local politicians joined the chorus of protest, and in 1970 the *Milwaukee Journal* labeled Wisconsin Electric "a highly privileged, government sheltered industry with a captive market."

Then came the energy crisis. The term had actually been in use since about 1972, when dwindling supplies of oil and natural gas were causing concern, but it became *the* crisis with the Arab oil embargo of 1973-74. Crude oil prices quadrupled within months, sending a shock wave through the American economy. The nation developed an instant energy awareness, evident in long lines at the gasoline pumps and general calls for conservation. In Milwaukee, city officials turned off the downtown Christmas lights, Miller Brewing canceled its tree-lighting ceremony, and stores shortened their hours.

The oil shortage boosted demand for other fuels, with predictable results. The price of coal nearly tripled between 1970 and 1975, intensifying the financial pressures on Wisconsin Electric. In the mid-1960s, when planning for Point Beach began, coal and nuclear fuel had been nearly identical in cost per kilowatt-hour. By 1975 coal was nearly three times more expensive, and the decision to build a nuclear plant seemed positively clairvoyant. Wisconsin Natural Gas faced a double problem: soaring prices and vanishing supplies. Consumers were using gas much faster than pipeline companies were finding new fields, and the life expectancy of known reserves was declining rapidly. The most severe blow came in 1975, when the subsidiary's supplier cut its allocation by 7.4 percent. Residential customers were protected but, after nearly doubling its sales in the previous 10 years, Wisconsin Natural was forced to impose a moratorium on new service to industries and other large consumers.

Fuel problems were only one consequence of the convulsions that followed the Arab oil embargo. Pushed upward by energy prices, the inflation rate spiked to 11 percent in 1974, its highest point since 1947, and moderated only slightly (to 9.1 percent) in 1975. For America's electric utilities, worse had come to worst. Construction budgets ballooned, interest rates soared, and labor and fuel costs kept on climbing. With its fixed costs rising out of control, Wisconsin Electric was forced to seek higher rates, but higher rates, coupled with conservation efforts, had an unanticipated outcome: a drop in demand. The system's output fell 1 percent in 1974 and continued to slide in 1975.

WEPCO's leaders were surprised, and more than a little nonplussed, to see their best projections turned upside-down by the marketplace. The utility was entering a vortex, a vicious downward spiral. As costs went up and demand went down, revenues were pinched severely, and the company was forced to seek yet another round of rate increases. In 1974 John Quale explained the necessity of a "make-whole" rate hike:

> *We support the national policy favoring conservation of all types of energy, but a reduction in electric consumption results in the need to recover the fixed charges on the large investment in electric facilities from the smaller amount of kilowatt-hours our customers use.*

Non-stop rate increases were, of course, hugely unpopular. After 50 years of steady decline, the average residential rate rose from 2.18 cents per kilowatt-hour in 1970 to 3.02 cents in 1975, a 38-percent jump. WEPCO pointed out that it was merely keeping up with the cost of living, which rose 39 percent during the same period, but consumers had come to rely on cheap electricity. They were especially upset at the prospect of paying more for less. As energy conservation became a national priority, only a handful of WEPCO's customers understood why their bills were going up while their consumption was going down. Rate hearings became outlets for pent-up anger, and one Milwaukee alderman blasted the utility as "unpatriotic and morally bankrupt."

Beating the drum for increased consumption was not an option. In the political climate of the 1970s, pushing energy usage was on a par with dumping hazardous waste in public parks. Predictably, Wisconsin Electric's promotional programs were among the first casualties of the energy crisis. The advertising budget, long a sore point with environmentalists, was slashed repeatedly. The Christmas Cooky Book, a seasonal staple since 1930, made its last appearance in 1973. The Sales Department was rechristened the Customer Services Department in 1974. In January, 1975, after nearly 70 years of continuous operation, the retail store in the Public Service Building closed its doors forever.

Although WEPCO was entering exceedingly dangerous waters, its leaders continued to sail into the wind. Plans for both the Koshkonong and the Pleasant Prairie Power Plants were unveiled in June, 1974, four months *after* the Arab oil embargo. Company experts fully expected demand to double every decade until the end of the century, and John Quale suggested that the energy crisis would create a "basically electric" American economy:

> *As other energy sources become ever more costly, and eventually extinct, the ability of the electric utility industry to use virtually any available fuel to produce energy will become increasingly important in our world.*

Quale and his colleagues believed that electricity was the key to "energy independence," the nation's newest buzzword. Sol Burstein went so far as to propose a "forced program" for the development of America's coal and uranium reserves. His emergency plan had three components: raising electric rates to finance new power plants, suspending

Customers lined up for the company's last Christmas Cooky Book in 1973. Its demise signaled the end of WEPCO's efforts to boost rather than limit energy consumption.

or deferring at least some environmental goals, and eliminating or restricting public participation in energy decisions. All three ideas ran directly counter to the dominant trends of the 1970s.

The system's executives, steeped in the ideology of progress, were reluctant to acknowledge that the rules had changed, that the old order had vanished. But the utility, like all American utilities, faced a chilling scenario. Spending billions on new power plants would undoubtedly raise electric rates. Higher rates would force consumers to reduce their energy usage. Reduced consumption would create surplus capacity and decreased revenues, without cutting the system's fixed costs. As those costs were spread across a shrinking revenue base, rates would rise again, further depressing demand, and so on ad infinitum. The void at the end of the tunnel was financial catastrophe. There was already evidence that the falling demand of 1974 was more than an aberration; new load studies predicted a drop in the utility's annual growth rate from 6.5 to 4.3 percent through 1980. As the energy crisis deepened, Wisconsin Electric looked ahead to a diminished future.

In the first weeks of 1975, a good deal earlier than most utility executives, John Quale experienced a conversion. Convinced that his company was trapped in a blind alley, he decided to follow a softer path. The first public sign of a shift in his thinking had surfaced in an October, 1974, speech to an investment group:

> *If the growth rate of electrical consumption could be cut in half over the next ten years from its historical rate, many of our Company's financial problems would be solved. A capital-intensive industry like our own just has no chance of showing financial and economic progress in a period of inflation such as we are now experiencing.*

But Quale was not yet ready to endorse conservation as the cornerstone of his company's future. That came in a report to shareholders in early 1975:

> *We must change. Wisconsin Electric Power Company is changing. We believe that our practices of yesterday were right for the times. Our practices of tomorrow must be right for the new times, the new conditions.... We must actively encourage wise and effective use of energy, eliminate waste of energy and thus conserve our energy resources. We expect to take an active role in reducing the growth rate in the demand for energy in our service area.*

Quale enlarged on the theme at the company's annual meeting in May, 1975:

> *The great growth patterns are no longer in the public interest. Nor are they possible. Energy shortages, inflation and the condition of the environment have created the need, not to stop growth altogether, but to bring a semblance of order to the growth that is inevitable and necessary.... Can energy conservation become a permanent ethic in America? We believe it must.*

Wisconsin Electric did not change overnight; Quale was dealing, after all, with an inertia developed over decades and a work force of 5,700 people, most of whom had begun their careers in more tranquil times. Slowly at first, the company edged away from the risks of headlong expansion to embrace the substantially new risks of a significant slowdown. The 1975-1979 capital budget was cut by a third. The construction deadlines for Koshkonong and Pleasant Prairie were pushed back. The utility announced plans for new load management techniques, new rate designs, and "a broad program of public education aimed at helping consumers cut their use of electricity."

John Quale did not live to see those plans become realities. He had been a vigorous, athletic man all his adult life, with a particular fondness for golf and curling. In August, 1975, however, he experienced a sudden illness during a golf outing, and a visit to the hospital produced a dread diagnosis: cancer of the brain, lungs, and bladder. Quale's case was too far advanced for effective treatment, and the disease claimed him within weeks. On September 8, shortly before his fifty-first birthday, Wisconsin Electric lost the leader who had guided the firm through some of the most harrowing years in its history.

The announcement of Quale's death was greeted with shock and disbelief in both the company and the community. WEPCO's future was suddenly an open question again, but the ship was by no means rudderless. In August, 1974, responding in part to the recommendations of Civil Defense planners, the board had adopted an "emergency succession plan." The directors had ranked the company's vice-presidents in order of succession, and the name at the top of the list was Charles McNeer's. On September 3, 1975, as John Quale lay on his deathbed, the board made McNeer president and chief executive officer.

The promotion was, to say the least, unexpected, but McNeer was well-prepared for the top job. One year younger than Quale, he brought a quarter-century of experience to the post, including six years in charge of WEPCO's operating side. In a speech to the system's managers a few weeks after his election, McNeer affirmed his own commitment to Quale's policies:

> *This company will devote its resources, its creativity and a significant amount of manpower and communication effort to encourage customers to conserve energy and to make the wisest use of the energy they do need to use. We will try hard to shift use of energy from "on-peak" to "off-peak" periods, but we will not promote the wasteful use of energy no matter what time of day it's used.*

McNeer closed the speech, his first as chief executive, with a plea for assistance:

> *I have no illusions about the magnitude of my job. I am responsible for management of a company that is to provide energy for two million people in a period of energy shortage.... I am going to need unqualified support from every man and woman in the company. So I ask you for that support, the same outstanding support you gave my predecessor. Together we can get the job done.*

John Quale's sudden death in 1975 threatened the company's equilibrium, but Charles McNeer, after six years at Quale's right hand, moved quickly to restore WEPCO's balance — no easy task in a time of pervasive uncertainty for America's utilities.

John Quale's death and Charles McNeer's promotion capped one of the most extraordinary, and most extraordinarily difficult, decades in Wisconsin Electric's history. Between 1965 and 1975, the system experienced multiple inversions: from an abundance of energy to shortages, from growth to relative stagnation, from a community pillar to a pilloried public enemy (again), from a prosperous calm to something approaching chaos. It is difficult to exaggerate the magnitude of the shifts. In 10 short years, Wisconsin Electric moved from stability to crisis to response. It was a rough ride by any standard, taking the company from smooth blacktop in daylight to a gravel road in the dark. As the decade closed, that road, with all its hazards, belonged to Charles McNeer.

Chapter 9

1975-1990

Breaking from the Pack

"The business of generating electricity," declared a Harvard utility expert in 1979, "has ceased to be a commercially viable enterprise." His statement, as flat and final as a death sentence, was rooted in some hard realities. It was clear, by the late 1970s, that too many utilities were locked in construction modes, adding capacity far beyond the current needs of their customers. The inevitable result was astronomical carrying costs, which forced rates up while growth in demand slowed down. As the widening gap between supply and demand drove prices still higher, the nation's electric utilities suffered both a loss of face and a loss of public faith. Consumers howled, regulators cracked down, and investors voted with their feet. By 1980 the average utility stock was trading for less than two-thirds of its book value.

Between 1975 and 1990, Wisconsin Electric proved the experts wrong. The company demonstrated that there was nothing inevitable about the utility crisis, and it showed that the business of generating electricity could be not only viable but vigorous. Responding creatively to new rules and changed conditions, WEPCO broke decisively from the pack. The company willed its own future, and the results, never guaranteed and sometimes in doubt, were an outstanding financial record and a world of opportunities that other utilities could barely imagine.

A Judicious Balance

Charles McNeer had a few things he wanted to get done after assuming the presidency in 1975. The list was fairly short: Change two million people from consumers to conservers, put the brakes on a billion-dollar building program, transform a passive public utility into a civic powerhouse, and reshape a corporate culture that had been generations in the making. As a product of the WEPCO system, McNeer had no interest in revolution, but his policies amounted to a 180-degree turn from past practice. The company's new commander began to spin the wheel as soon as he took the helm.

McNeer did not match the stereotype of the hard-charging iconoclast. Low-key, soft-spoken, and deeply private, he seemed, at first glance, more a chief engineer than a chief executive. Beneath the composed exterior, however, lay a visionary's intelligence, an appetite for decision-making, and a determination that bordered on the absolute. McNeer knew what he wanted. He had little tolerance for wasted effort and even less for indecision; his subordinates learned to come to the point quickly and clearly. But McNeer's autocratic tendencies were tempered by his relentless logic and his basic flexibility. What he wanted was options, and the details of WEPCO's strategy

Departing from past practice and conventional wisdom, Charles McNeer changed Wisconsin Electric's priorities more profoundly than any president since James Mortimer.

changed constantly during his tenure.

Wisconsin Electric was not, of course, a blank slate when McNeer took office. John Quale had called the turn to conservation in the months before his death, and Wisconsin's Public Service Commission, one of the most progressive in the nation, applied constant pressure to slow the growth in demand. Nor was McNeer a one-man show. He depended on a stable core of seasoned executives, and chief among them were Sol Burstein and Russell Britt. Burstein had become a board member and WEPCO's first executive vice-president at the end of 1973. He was one of the most powerful men in the company, but power plants remained his chief concern. Burstein's technical expertise, particularly in the nuclear field, earned him an international reputation. Even the critics were impressed. David Comey, one of the anti-nuclear crusade's scientific leaders, offered surprising praise: "If everyone ran their nuclear power plants like Sol Burstein runs Point Beach, there wouldn't be much opposition to nuclear power."

Sol Burstein remained one of the most influential figures in the company, with sole responsibility for WEPCO's power plants.

Russ Britt rose even higher in the corporate hierarchy. A native of Madison with degrees in both accounting and engineering, Britt had joined WEPCO in 1948, two years before McNeer, and had risen, over the years, to the top of the system's financial side. In 1975 he won election to the board of directors, succeeding John Quale, and joined Sol Burstein as an executive vice-president. In 1982, when McNeer moved to the chairman's post, Britt became WEPCO's president and chief operating officer. The two executives had vastly different temperaments — Britt was aggressive, even brash, where McNeer was temperate and cerebral — but they formed an effective team. "I ran the real world," said Britt, "and Charlie ran the maybe world."

Other figures and other forces aided in WEPCO's transformation, but it was Charles McNeer who brought it off. No one since James Mortimer, who moved the company's emphasis from traction to power during the World War I era, had changed the system's priorities so profoundly. And what were McNeer's priorities? He had one overriding goal: to find a judicious

Russell Britt joined Wisconsin Electric fresh out of college in 1948. By 1974 (above) he was well on his way to becoming the system's second-in-command.

In March, 1976, only six months after McNeer took office, the worst ice storm in company history downed hundreds of poles and power lines. It took 400 crews nearly 10 days to restore service.

balance between supply and demand. In order to keep costs at a reasonable level, McNeer tried to build as little as possible, and the best way to minimize construction was to minimize demand. Wisconsin Electric launched dozens of programs aimed at curbing the region's appetite for power. McNeer discovered, in effect, a third fuel. "We must meet," he said in 1976, "the bulk of the requirements of the people we serve with a combination of conservation, coal and uranium." Every kilowatt-hour deferred or eliminated was one less that had to be generated.

It should be stressed that Charles McNeer did not share the no-growth philosophy of the most ardent environmentalists. Although he was an avid bird-watcher, he was not a tree-hugger. In 1975, for instance, McNeer urged consideration of the "total environment":

> *I see the inadequate housing of the poor as part of the environment. I see jobless workers as part of the environment. I see as part of the environment the young people of our high schools and colleges, people who will soon be entering the job market....*
>
> *Without expanding supplies of reliable, affordable energy, there is no way the total environment of our community can improve.*

Both sides of the supply-demand equation were equally important to McNeer. Growth, in his view, was inevitable as well as desirable; he had no interest in putting the region, much less his own company, on a starvation diet. McNeer's chief concern was the *rate* of growth. Wisconsin Electric sought to build as little as possible, but always more than enough. Alternating deftly between the accelerator and the brake, WEPCO's chief executive tried to find the precise point of balance between the public's need for power and the company's need for profits. Social and environmental concerns remained paramount, but every new program, every new building project was judged finally on one criterion: whether or not it made *economic sense.* It was a principled pragmatism, not an ascetic idealism, that shaped the policies of the McNeer era.

Those policies appear to be, in hindsight, the epitome of common sense. What could be more rational than trimming both supply and demand to manageable proportions? But Wisconsin Electric was one of only a handful of American utilities that embraced conservation as a corporate strategy. In 1969 a *Fortune* reporter had dismissed the utility industry as "clumsy" and "sluggish," a field dominated by "generally unimaginative men, grown complacent on private monopoly and regulated profits." Events of the 1970s bore out his criticism. Most utility executives were convinced that inflation and recession were temporary aberrations, and that the great growth patterns of the postwar period would return in full force. They were wrong, of course, and the result was surplus capacity approaching 50 percent in some regions. But there was no guarantee that McNeer was right. A rapid return to "normal" economic conditions, or a failure of the company's conservation efforts, might have left WEPCO in the lurch, and a sudden downturn might have idled some very expensive plants. The key was a judicious balance between supply and demand.

Despite vigorous efforts to spread information and build support, Wisconsin Electric and its partners were forced to drop their plans for a nuclear plant at Koshkonong.

The Supply Side

Between 1965 and 1975, despite the lengthy delays and acrimonious debates, Wisconsin Electric's supply side had grown just fast enough to keep pace with the demands of its customers. The system's capacity had increased 68 percent, while its peak load had risen 67 percent. Although demand studies predicted slower growth in the future, the utility considered it imperative, at first, to continue the ambitious construction program of the Gruhl and Quale years.

The Koshkonong Nuclear Plant was at the top of the agenda when Charles McNeer took office. WEPCO and its partners were still awaiting word from state authorities at the time of John Quale's death, and the word, when it came, was not good. In November, 1976, the Department of Natural Resources declared the project "environmentally unacceptable." The Rock River was hardly a pristine trout stream, but the DNR concluded that the Koshkonong plant would use "excessive amounts" of river water and return it in a condition below prevailing standards. In the same month, the Public Service Commission ordered Madison Gas & Electric, a utility that was experiencing financial problems, to dispose of its 6.3-percent stake in the Koshkonong project. A reporter for *The Nation* called the double rulings "perhaps the largest victory in the nationwide movement to halt the perilous rush to nuclear power."

The remaining partners — Wisconsin Electric, Wisconsin Power & Light, and Wisconsin Public Service — were by no means ready to concede defeat. Although the project's budget had ballooned from $965 million in 1974

An alternate site emerged in 1977: a former military installation near the tiny settlement of Haven in Sheboygan County.

to $1.4 billion in 1976, the utilities continued to believe that the new plant's 1,800 megawatts were vital to their collective future. The Koshkonong site, however, was quietly dropped. "Even though we think we would ultimately prevail at Koshkonong," said Charles McNeer in 1977, "it may take years to contest the point. We don't have time to fight legal battles when there's a chance we'll run short of electricity for our customers." With the tacit approval of state officials, the partners turned their attention to a Lake Michigan site near the town of Haven, a few miles north of Sheboygan.

The change of venue was more than a little exasperating. Planning, always an inexact science, was practically impossible in the 1970s, a point the Haven partners underscored in a 1977 report:

> *The inflationary economy, followed by a prolonged recession, and extensive environmental legislation, coupled with uncertain air and water quality criteria, continuing debate concerning nuclear power on national and local levels, a lack of federal, state and local regulatory agency coordination, along with the increased and uncertain time required for the regulatory process, have combined to make the planning process an extremely frustrating task.*

Wisconsin's regulators made no attempt to ease the Haven group's frustrations. The Public Service Commission notified the partners that it would not consider more than one nuclear reactor on the new site, forcing them to cut the plant's capacity in half. (The budget for even half a plant climbed to more than $1 billion.) In August, 1978, the utilities filed a formal request to build a 900-megawatt nuclear powerhouse at Haven, and the round of site studies and environmental impact statements began all over again.

Then came Three Mile Island. On March 28, 1979, a chain of human missteps and mechanical failures crippled a nuclear plant on the Susquehanna River near Harrisburg, Pennsylvania. A reactor core suffered irreparable damage, and fears of a full-scale meltdown prompted a partial evacuation of the surrounding area. Coming on the heels of *The China Syndrome,* Hollywood's version of a nuclear near-miss, the TMI "incident" seemed to be a case of life imitating art. The media had a field day. With the possible exception of Chernobyl, which suffered (and caused) much more serious damage in 1986, Three Mile

Island became the best-known nuclear plant in the world.

TMI's impact on public health was, fortunately, negligible. The real victim of the fallout was America's nuclear power industry. Responding to a panic-stricken public, federal regulators tightened licensing procedures, mandated more rigorous training programs, and imposed a host of new safety standards. The work force at WEPCO's Point Beach plant nearly tripled in 10 years as a direct result of post-TMI safeguards. In the prevailing political climate, the Haven Nuclear Plant was dead, and Wisconsin Electric knew it. "It was obvious to me," said Charles McNeer years later. "That's why we stopped our planning and reduced our expenditures to the minimum possible until we were able to gracefully get out of the plans." On February 29, 1980, a leap day, WEPCO and its partners notified the PSC that they were withdrawing their request to build a nuclear plant at Haven.

Public fallout from the Three Mile Island incident threatened the future of America's nuclear power industry. (Milwaukee Public Library)

No one at Wisconsin Electric sent thank-you notes to state officials, but there is little doubt that the multiple delays in the Koshkonong/Haven project were a stroke of pure luck. The delays were undeniably expensive; WEPCO and its partners spent $50 million on the plant without turning over a single shovelful of earth. (The pre-building costs for Point Beach, by contrast, were $4.4 million.) If, however, the project had proceeded as smoothly as Point Beach, the utilities might have been at the point of no return when Three Mile Island failed. Utilities with plants under way in 1979 found their costs multiplying exponentially and their public support vanishing, with predictably disastrous results. The budget for Seabrook 1, a 1,150-megawatt nuclear plant in New Hampshire, swelled to $4.9 *billion* by 1987, driving the project's managing partner into bankruptcy.

Wisconsin Electric was spared the agony of Seabrook, but dodging a bullet did not solve the company's supply problems. WEPCO's major concern was the growth in peak demand, which threatened to outstrip the system's capacity for a few weeks each summer. In earlier decades, the utility's peak output had always occurred in December, when days were short, stores were open late for Christmas, and streetcars were filled with shoppers. The growth of air conditioning, coupled with the demise of electrified transit, altered the balance of power after World War II. Year by year, WEPCO's summertime output rose steadily against the winter peak, finally overtaking it in 1968.

The growing seasonal imbalance created a pressing need for peaking plants. In 1975, five years before the demise of Haven, the utility announced plans for a 213-megawatt combustion turbine plant in Germantown, just northwest of Milwaukee. Regulatory approval was not automatic, particularly after Germantown residents organized to fight the project, but the case took on a new urgency in 1977, when a turbine explosion disabled a 130-megawatt unit at Oak Creek. Persuaded that WEPCO needed new capacity immediately, the PSC put the Germantown turbines on the fast track, and they were ready for the dog days of the summer of 1978. Fired by oil, the peaking plant was nearly 10 times more expensive to operate than Point Beach; Wisconsin Electric used it as sparingly as possible.

The Pleasant Prairie plant, on the drawing board since 1974, was another holdover from the Quale years. Convinced that the 1,160-megawatt plant would raise ozone pollution to unacceptable levels, the Department of Natural Resources tried to block the project in 1976. Convinced that the plant was absolutely essential to its

With the nuclear option nullified, Wisconsin Electric found more conventional solutions to its capacity problems. The Germantown peaking plant, fueled by oil, was finished in 1978.

The first unit of the coal-fired Pleasant Prairie Power Plant went on line in 1980. Known as "P4" to most employees, it quickly became a star of the system.

future, Wisconsin Electric took the DNR to court. The judges ruled in favor of WEPCO, and work on the Pleasant Prairie Power Plant (known within the system as "P4") began in the summer of 1976. Delayed by harsh weather, a shortage of skilled workers, and a strike, the first 580-megawatt unit finally began commercial operation on June 30, 1980. The project had required a total of 49 permits — 30 from the state, 11 from the federal government, and 8 from local authorities — and it had cost $500 per kilowatt, well above the $345 budgeted in 1974.

Pleasant Prairie was the system's first new coal plant since the opening of Valley in 1969, and the completion of its initial unit allowed WEPCO to retire two old powerhouses. For all their historical importance, East Wells and Lakeside had long since outlived their usefulness. Pollution problems limited their output, and it was practically impossible to find spare parts for turbines built in the early 1900s. East Wells, the birthplace of pulverized-coal technology, was shut down permanently in 1982. In a well-publicized gesture of civic munificence, the company donated the building to the Milwaukee Repertory Theater in 1984. After a top-to-bottom remodeling, the plant reopened, fittingly, as the Powerhouse Theater. The old landmark provided a one-of-a-kind home for the region's major theater company, and it also served as the cornerstone for a new landmark: the Milwaukee Center, a hotel/office/retail complex that rose in the shadow of City Hall. The Lakeside plant, another engineering shrine, was decommissioned in 1983 and sold to a group of Illinois investors. The partners razed the powerhouse, but their plans for condominiums, stores, and a marina on the lakefront languished for years. It was not until 1995, when Harnischfeger Industries broke ground for its corporate headquarters, that the site showed tangible signs of new life.

There was surprising beauty in the skeleton of a Pleasant Prairie cooling tower ...

Work at Pleasant Prairie continued as the old plants were retired. Completion of the second unit, originally scheduled for 1980, was postponed repeatedly, not because of regulatory delays but because the company felt no need for its capacity. Growth in demand was slowing to a crawl, and WEPCO held off until the last possible moment. The plant was finally finished on July 1, 1985, near the end of a sharp inflationary cycle. Although Pleasant Prairie's second unit was virtually identical to the first, it cost a total of $600 per kilowatt — 20 percent more than its twin. "P4" quickly displaced Oak Creek at the top of the coal pile. The plant was both more efficient and more reliable than its older cousin, and its operating costs rivaled those of Point Beach.

Point Beach itself had long been one of the most reliable nuclear plants in the world. This standout was not, however, without its problems. In the mid-1970s, the plant's steam generator tubes, which transferred heat from the reactor loop to the turbine system, began to leak alarmingly — most likely as the result of corrosion from water-treatment chemicals. Leaking tubes forced the periodic shutdown of both units for repairs, but stopgap measures failed to solve the problem. In 1983 and 1984, Wisconsin Electric replaced the first unit's steam generators entirely and inserted sleeves in the second unit's tubes — at a total cost of $65 million. Restored to full-time operation, Point Beach continued to earn its keep as a mainstay of the system.

Although its supply side rested firmly on a foundation of coal and uranium, the company showed a willingness to explore some less conventional energy options during the McNeer years. Wisconsin Electric funded research into such futuristic technologies as nuclear fusion and superconductivity, and the utility established its own demonstration projects in solar power, wind power, and stored cooling. With a change in federal law in 1978, WEPCO actively encouraged the growth of alternative energy producers, who sold their power to the company at market rates. By 1986 there were 60 independent producers in the service area. Most operated windmills with capacities of less than 10 kilowatts, but the group also included Waste Management, the region's major trash collector. Using methane from the decomposing garbage in its landfills, the company operated turbines with a total capacity of 10 megawatts.

Garbage was also the key ingredient in the utility's most novel energy experiment. In 1977, after three years

... and a touch of sadness in the passing of the Lakeside Power Plant.

WEPCO conducted experiments with both wind and solar power at a Point Beach demonstration facility.

of preliminary work, Wisconsin Electric and the American Can Company launched a project called Americology. Most of Milwaukee's garbage was hauled to a plant in the Menomonee Valley, where it was sorted, shredded, and then transferred to Oak Creek, where the burnables were "air-classified" (fluffed) and blown into the boilers of units 7 and 8. "Trash power" made sense both ecologically and economically. WEPCO hoped to burn 150,000 tons a year, saving precious landfill space and reducing coal consumption in the two units by 10 percent. In practical terms, however, the Americology project was a disaster. Not only did smoke from the burning garbage regularly exceed emission limits, but glass in the gumbo fused to the boiler walls, forming a thick layer of slag. When shotguns failed to remove the deposits, Wisconsin Electric was forced to call in a team of dynamiters from Missouri. The project was quietly terminated in 1980.

However marginally, supply-side experiments like Americology and methane turbines helped to forestall the need for new power plants. In 1985, as Pleasant Prairie's second unit neared completion, Wisconsin Electric called it "the last major plant of the twentieth century." Although the company made substantial additions to its supply side after 1975, the real story was the plants it *didn't* build. As some projects, notably Koshkonong/Haven, were canceled and others were postponed, WEPCO departed abruptly from its long-term expansion program. In the 15 years between 1970 and 1985, the utility added 2,500 megawatts of capacity. In the 15 years from 1985 to 2000, the company planned to build a grand total of 300 megawatts more.

The Americology project was a valiant attempt to turn Milwaukee's garbage into fuel for the Oak Creek Power Plant.

Unfortunately, the trash left such thick deposits on the boiler walls that not even shotgun blasts could remove them.

The dramatic slowdown, which saved roughly $1.5 billion in construction costs, was the direct result of changes taking place on the other side of the power equation: the demand side.

The Demand Side

Even a scaled-down construction program was expensive. Pleasant Prairie alone cost $638 million, and the replacement or improvement of substations, transmission lines, and other facilities required millions more. Wisconsin Electric's construction budget peaked at $294 million in 1980, much of it borrowed during a period of record-high interest rates. When the company tried to place an $80 million bond issue in 1980, no underwriter would take the bonds for less than 14 percent. Russell Britt and his colleagues were so disgusted that they took the highly unusual step of rejecting every bid. Two months later, unable to generate the cash internally, the company finally sold the bonds — at 13.92 percent. WEPCO's long-term debt approached $800 million by the end of 1980, well above the $460 million figure for 1975, and the company's annual interest charges climbed past $70 million.

The utility was in no financial danger. Wisconsin Electric had been a stronghold of conservative accounting since the days of John I. Beggs; its dividend payouts, construction allowances, and depreciation reserves were among the most judicious in the industry. The company, in other words, paid cash when it could, and its debt ratio actually improved after 1975, never exceeding 52 percent of

Non-stop rate hikes drew crowds of protesters to public hearings. Richard Abdoo, shown above at a 1981 encounter, was the company's designated "punching bag."

capitalization. But the unrelenting pressures of inflation, which spiked to 13.5 percent in 1980, forced a continuation of the rate increases that had been going on without interruption since 1968. The average residential rate more than doubled between 1975 and 1983, climbing from 3.02 cents per kilowatt-hour to 7.05 cents. As the flood of rate requests from WEPCO and other utilities threatened to overwhelm the Public Service Commission, "regulatory lag" became an issue. In 1981 the company began to file new requests before its older cases were settled, a practice derided by consumer groups as "pancaking."

Wisconsin Electric took pains to point out that its rates remained among the lowest in the country, but consumers found a steady diet of increases hard to stomach. Rate hearings were forums for the disaffected, and the utility's officers heard their company pummeled as "unthinking," "uncaring," and "inhumane." Richard Abdoo, who joined the system in 1975 as director of planning, was WEPCO's designated spokesman at most hearings. In 1981 a *Milwaukee Sentinel* reporter described Abdoo as "Wisconsin Electric's punching bag, the man who is cornered, yelled at and thumped with words at dozens of public meetings." The company's executives took their lumps in private as well. When Russell Britt became chief operating officer in 1982, he said, "If at the end of five years I can go to a party and not have to spend half the evening defending my company, I will feel successful."

Non-stop rate hikes did not win the company any friends, but they were, in fact, half the battle in WEPCO's efforts to hold down its growth rate. The utility's planners became expert at gauging the "price elasticity of demand," a ten-dollar term for the tendency of consumers to use less of a product as its price increased. Appeals to patriotism, guilt, and the well-being of the planet helped to reduce energy consumption but, as always, it was appeals to the pocketbook that counted most.

The other half of the demand-side battle was a barrage of programs designed to reduce the region's appetite for energy. After decades of assiduous attempts to sell electricity, some veterans found the company's about-face more than a little jarring, but Wisconsin Electric became a national trailblazer in efforts to sell conservation. The most visible activity was a broad program of public education. WEPCO's advertisements, which had touted the virtues of "All-Electric Living" only a few years earlier, urged consumers to "Put your house on a diet." In place of Christmas

Cooky Books, the company published booklets on energy conservation, giving away 60,000 before the end of 1976. Workers in the old Home Service Bureau began new careers as "conservation counselors," and the focus of their classes shifted from cooking to caulking. The Energy Facts Phone, a consumer call-in service with recorded messages on 75 topics, debuted in 1980 and attracted 16,000 callers in its first three months. A troupe of itinerant speakers spread the gospel of conservation to civic and community groups, and Ernie the Energy Rabbit made the rounds of the region's schools. Other activities ranged from the Energy House, a showcase for the latest in weatherization techniques, to appearances by Offalot, the company's "energy mascot."

It did not take long for consumers to get the message. Wisconsin Electric resumed its annual survey of customers' attitudes and behaviors in 1976. Within a year, the proportion of customers who understood the concept of peak demand jumped from 40 to 70 percent. Within two years, the number of residents who did their laundry at night climbed from 17 percent of the population to 42 percent, and the number who ran their dishwashers in the evening rose from 62 to 81 percent.

The sweeping educational campaign was clearly changing behavior, but WEPCO also developed specific programs with specific targets. In 1978, after a lengthy test in the Belgium-Cedar Grove area, the utility began to recruit customers willing to have their water heaters shut off during the hours of peak demand for electricity, which happened to be the hours of lowest demand for hot water. For a credit of $1.50 (later $4.50) each month, customers agreed to have their heaters equipped with remote-control switches operated from the System Control Center in Pewaukee. Participation rose steadily, from 40,000 households in 1980 to 85,000 in 1982 and 100,000 (the final target) in 1986. The water-heater program saved Wisconsin Electric 60 megawatts in "avoided capacity," and customers found that there was still plenty of hot water for morning showers.

Time-of-use rates were another 1978 innovation. The new rates reflected the actual cost of producing electricity, which was highest, of course, during the hours of peak demand. All power plants are not created equal; their operating costs vary wildly as a function of age, fuel, and general efficiency. Base-load plants are the heartbeat of every utility, but peak demand

Higher rates were an effective "stick," but WEPCO favored carrots in its campaign to reduce demand. Ernie the Energy Rabbit reached out to younger audiences, while other programs addressed their parents' concerns.

brings less-efficient plants into service, and less efficient plants are invariably more expensive to operate. In 1975, for instance, WEPCO's costs per kilowatt-hour ranged from $0.47 at Point Beach to $1.09 at Oak Creek, $1.59 at Port Washington, and $2.92 in the oil-fired turbine plants. As daily demand reached its peak, the overall cost of production multiplied, and time-of-use customers paid accordingly; day rates were nearly five times higher than night rates. The Public Service Commission made the new schedule mandatory for the system's largest residential customers and for all commercial and industrial accounts. (The region's foundries promptly moved their melting to the night shift.) In 1981 the rates were opened to all households on a voluntary basis, and nearly 15,000 customers joined the program. By 1983 half of all Wisconsin Electric's sales were billed on a time-of-use basis.

Seasonal pricing, a variant of time-of-use rates, affected everyone. As America became an air-conditioned society, WEPCO's warm-weather peaks were dependably higher than those of the winter months, with a predictable impact on operating costs. When the utility asked for a rate schedule that reflected the seasonal variance, the PSC obliged, and then some. In 1978 the Commission made summer rates 50 percent higher than those charged during the rest of the year — a far wider gap than the company had requested. The result was a public relations headache that approached migraine proportions. WEPCO's telephone operators were soon fielding nearly 700 calls a day from angry customers. Those who sweltered through the dog days without air conditioning were especially irate. Under steady pressure from Wisconsin Electric, the PSC gradually reduced the summer-winter differential: from 50 percent in 1978 to 30 percent in 1980, 20 percent in 1981, and 5 percent in 1983. To the immense relief of company officials, summer rates were finally eliminated in 1986.

Another conservation program won universal praise: free energy audits. In 1976 retrained sales engineers began to make calls on the system's largest commercial and industrial customers, studying their energy practices and suggesting improvements. Wisconsin Natural Gas, WEPCO's largest subsidiary, extended the service to residential customers in 1978, restoring, on at least a one-time basis, the personal contact of the pre-computer years. Gas was still in short supply when the program began, making conservation a high priority. After 40 hours of intensive training, Wisconsin Natural's energy auditors, many of them college students, made two-hour visits to homes throughout the company's service area. Within two years, more than 13,000 homeowners had been tutored in the mysteries of R values, air infiltration, and solar gain.

With the discovery of new gasfields, Wisconsin Natural's supply problems eased; the 1975 moratorium on service to new industrial and commercial customers was lifted in 1979. The residential market continued to show the steadiest growth. Despite a round of rate increases, gas remained significantly cheaper than fuel oil, and thousands of households (10,289 in 1979 alone) switched from oil to gas heat. Although the need for conservation became somewhat less acute, energy audits and other programs remained a central focus at Wisconsin Natural.

Wisconsin Electric began to offer home energy audits in 1981, three years later than its gas subsidiary. By the end of 1983, the two companies had inspected a total of 100,000 homes from top to bottom. Every participating household received a free "energy kit," with an assortment of weatherstripping, rope caulk, duct tape, and booklets on conservation. For consumers who really wanted to put their houses "on a diet," the companies offered weatherization loans and substantial discounts on water heater blankets, caulking, pipe insulation, and other "Wrap Up/Seal Up" supplies. Response to the program was resoundingly positive.

By 1985 the WEPCO system offered no fewer than 27 different energy conservation programs. Not all were runaway successes. Response to a solar water-heating promotion was lukewarm at best, largely because of the costs involved, and only a handful of industrial customers showed any interest in lower rates for interruptible service. But there were substantially more hits than misses. Results on the gas side of the business were little short of spectacular. Between 1975 and 1985, Wisconsin Natural added 46,537 customers to its lines, an increase of 26 percent, but the system's total sales moved in the opposite direction, falling 11 percent in the same decade. Average gas consumption per customer actually decreased at a rate of 3 percent per year. Electric customers were nearly as responsive. WEPCO's forecasts of annual demand growth dropped as steadily as a thermometer in January: from 4.3 percent in 1975 to 3.3 percent in 1978, 2.8 percent in 1980, and 2 percent in 1982. The forecasts proved to be conservative. Under the double pressure of higher prices and aggressive conservation efforts, the utility's growth curve flattened. Between 1975 and 1985, Wisconsin Electric's output grew at an average rate of only 1.7 percent annually — less than one-third the rate of the previous decade. The gospel of conservation had found its believers.

Reaping the Rewards

By 1983, the approximate midpoint of Charles McNeer's term as chief executive, it was clear that the war was being won. Supply and demand were in balance. The company's construction program had been pronounced "manageable." Long-term debt was on the decline, and net income, after falling behind inflation in the late 1970s, was growing at a double-digit pace. While its counterparts across the country were mired in debt, debate, and distractions, Wisconsin Electric entered a period of enviable stability. In 1984 WEPCO became the only major utility in the nation to *lower* its rates, and the company repeated the trick in 1985. The Milwaukee area boasted the eighth-lowest electric rates in urban America.

The applause was soon audible. The company's approval rating, one of the most closely-watched variables in the annual consumer poll, soared from 65 percent in 1981 (the year of an 18-percent rate hike) to 79 percent in 1984 and 88 percent in 1985. Investors were just as enthusiastic as customers. After falling to a low of $12.75 per share in 1980, Wisconsin Electric's common stock soared to $28.50 in 1983, more than 60 percent higher than the American utility average. In 1984 Moody's Investors Service upgraded the company's bonds to AAA — the highest rating possible.

WEPCO was also turning a few heads within the utility industry. In 1982 *Electric Light and Power,* a respected trade journal, named Wisconsin Electric its "Utility of the Year." The magazine's editors cited the company's "strong financial performance ... brought about by early recognition of changing growth trends in demand and the advantages of adjusting its construction program in light of those trends." A 1983 poll of Wall Street analysts ranked WEPCO second among the best-managed utilities in the country,

Volume 60, Number 11
November 1982

Electric Light & Power

THE NEWS MAGAZINE OF ELECTRIC UTILITY MANAGEMENT AND TECHNOLOGY

WEPCo named Electric Utility of Year

BARRINGTON, IL—Wisconsin Electric Power Co., which serves three territories in Wisconsin and the Upper Peninsula of Michigan, has been chosen Electric Utility of the Year for 1982. The announcement was made by *Electric Light & Power* magazine Editor-in-Chief Robert A. Lincicome.

The Milwaukee-headquartered

lysts as "difficult" were important factors in company management's ability to gain better control of its own future prospects than many other utilities are capable.

WEPCo has curtailed its construction program to the point where it anticipates no new additions until 1992

beyond the three projects now being finished. As a result, the company believes it will be able to meet 93 percent of its construction fund requirements, and 69 percent of total capital needs, from internal generation.

WEPCo's heavy emphasis on two-way communication with its publics—

customers, shareholders and employees—also was rated heavily in the decision, along with the company's two-year budgeting procedure that allows rate increases to begin with the start of the target year, instead of part way through the year.

The Wisconsin utility, which serves some 2-million people in its three operating territories in the industrial-

Wisconsin Electric's strong performance soon earned the applause of industry experts.

and a 1985 study by the National Association of Utility Commissioners ranked the company first in financial performance over the previous decade. Charles McNeer himself was singled out for praise, earning awards for management excellence from both *Industry Week* and *The Wall Street Transcript* in 1985.

The tributes and testimonials of the mid-1980s were gratifying, but McNeer still had a business to run. Operating from a position of unusual strength, he launched new initiatives on both the supply and the demand sides of the business. There were familiar pressures on the supply side. Slow growth is not the same as no growth, and every new kilowatt-hour of demand diminished the system's aging reserve capacity. With the exceptions of Pleasant Prairie and Germantown, all of the company's plants were at least middle-aged by the 1980s, and some were practically senior citizens. Building new facilities, however, was not an attractive option. Wisconsin Electric adopted a "least-cost" approach to its supply problems, which meant, in practice, updating older plants rather than replacing them. Like a homeowner choosing to remodel rather than move, WEPCO made detailed plans to take its plants into the twenty-first century.

In December, 1985, the company announced a $600 million program with two major goals: to modernize the Port Washington and Oak Creek plants and, at the same time, to drastically reduce the sulfur-dioxide emissions from both. The plan called for the replacement of boiler drums and other worn-out equipment, but the heart of the program was a new technology: fluidized-bed combustion. Developed in Europe, FBC systems burned a blend of powdered coal and limestone on beds placed over a boiler's water tubes. The blend burned at lower temperatures than pulverized coal, prolonging the life of the boiler and, better yet, the limestone absorbed 90 percent of the SO_2 released during combustion, converting it to gypsum. Charles McNeer pronounced the FBC system "the boiler of the future," a technical and environmental breakthrough. Oak Creek's well-used north units, built between 1953 and 1957, were slated for conversion by 1992. Plans for a pressurized version of the system in Port Washington's fifth unit (built in 1950) hinged on a "clean coal" demonstration grant from the Department of Energy.

Two new developments gave the utility time to consider its options more carefully. In 1980, when Wisconsin Power & Light began work on a 380-megawatt addition to its Edgewater plant in Sheboygan, Wisconsin Electric purchased a 25-percent interest in the project. The $306 million unit finally came on line in 1985, increasing WEPCO's capacity by 95 megawatts. (Although no one noticed, the arrangement gave Wisconsin Electric the Sheboygan presence John I. Beggs had so fervently sought sixty years earlier.) The utility made a substantially larger purchase in 1987. Nearly a decade earlier, McNeer and his colleagues had seriously considered buying Upper Peninsula Power, a utility serving the territory north of their system's holdings in Upper Michigan. The deal fell through when closer study failed to reveal "enough potential advantages." In 1987, however, WEPCO learned that the major source of UP Power's electricity — the Presque Isle coal plant — was for sale. Perched on the Lake Superior shore at Marquette, Presque Isle was a 596-megawatt powerhouse built in nine stages between 1955 and 1979.

The Presque Isle plant, built to serve the iron mines of Upper Michigan, offered a ready-made answer to questions of supply.

Completion of the Pleasant Prairie Power Plant in 1985 gave the system a new workhorse.

Wisconsin Electric bought the facility for $248 million, or $416 per kilowatt — less than half the cost of a comparable new plant and substantially less than renovation of an old one. Presque Isle had been built to serve nearby iron mines, and one of those mines, the Empire, displaced Kenosha's Chrysler plant as the system's largest customer; the Empire and its neighbor, the Tilden, soon accounted for 10 percent of WEPCO's load. The purchase also extended Wisconsin Electric's reach from Lake Michigan to Lake Superior.

The addition of 95 megawatts from Edgewater and 596 from Presque Isle made some of the company's older plants, and newer plans, expendable. The Commerce powerhouse, after years on standby status, was formally retired in 1988, and the four old units at Oak Creek North followed in the next two years. Those units were to have housed the largest fluidized-bed combustion installation in the world. With their retirement and the rejection of WEPCO's "clean coal" application for Port Washington's fifth unit, the FBC system became the boiler of the more distant future. The company proceeded with a general refurbishing of Port Washington. Federal authorities, however, ruled that the project was a "modification," altering the plant so extensively that it would have to meet the emission standards applied to new facilities. Although WEPCO officials took strenuous exception to the ruling, they shut down Port's fifth unit permanently in 1990 to keep the plant in compliance.

As some older plants were retired and others were shut down for remodeling, Pleasant Prairie and Point Beach came to new prominence as the system's workhorses. Pleasant Prairie remained a pleasant surprise, setting new standards of efficiency and reliability for the system's coal plants. Even its fly ash, once a troublesome waste product, became an asset. The quality of the ash from "P4" made it a natural concrete additive, and the material found a permanent place in the Milwaukee Center, the 100 East Building, the Bradley Center, and other Milwaukee landmarks that rose during the building boom of the 1980s. WEPCO's fly ash sales soared from 18,000 tons in 1980 to 250,000 in 1989, and Pleasant Prairie was the system's leading supplier by a wide margin. Customers called from as far away as Minneapolis and Pittsburgh.

Point Beach reached an important milestone in 1986 — 100 billion kilowatt-hours of generation since 1970 — and the plant continued to anchor the system. In its milestone year, the powerhouse represented only 21 percent of WEPCO's capacity but generated 37 percent of its electricity. Point Beach, however, had a waste problem that proved a good deal more difficult to solve than Pleasant Prairie's: a growing collection of spent fuel assemblies. In the early 1970s, WEPCO had shipped the radioactive rods to facilities in Illinois and New York for reprocessing, but concerns about public safety and possible diversion kept reprocessing on hold. (The plants could have produced weapons-grade plutonium.) In the tidal wave of anti-nuclear sentiment that engulfed the industry after Three Mile Island, Wisconsin Electric was forced to haul the assemblies back to Point Beach. America's utilities continued to look to the federal government for a long-term solution to the nuclear waste problem, but indecision on the federal level hardened to a state of paralysis. By 1989, with the storage pools at Point Beach nearing capacity, WEPCO developed plans to store its spent fuel rods in "dry casks" on the site.

Wisconsin Natural Gas solved some supply-side problems of its own in the late 1980s. The subsidiary had been dependent on one supplier since it entered the business in 1950. With deregulation of the gas industry in 1985, pipelines proliferated. Wisconsin Natural made connections with a second supplier in October, 1988, and a third only days later. Deregulation

Good Cents awards turned the old Medallion Home program on its head, rewarding conservation rather than consumption.

created other pressures, but access to three pipelines enabled the company to do a better job for its customers. "Comparison shopping" generated a savings of $11 million in 1988 alone.

In contrast to the flurry of additions, subtractions, and modifications on the supply side in the later 1980s, activities on the system's demand side were relatively straightforward. A few new programs were added to the conservation campaign. In 1985 Wisconsin Electric began to honor energy efficiency in new construction with Good Cents Home awards — a complete reversal of the old Medallion Home program — and in 1987 the company launched a marketing blitz for compact fluorescent light bulbs, which used 75 percent less power than incandescents. The centerpiece of demand-side efforts, however, was undoubtedly the Smart Money Program. Unfurled in 1987, Smart Money was a $73 million umbrella for a host of programs designed to shave 130 megawatts of demand over a two-year period. Russell Britt announced Smart Money as "the biggest single commitment to energy conservation and efficiency in the history of the electric power industry."

Wisconsin Electric had used both the carrot and the stick in its earlier conservation programs. Water heater shut-offs and energy audits provided carrots, and time-of-use rates and seasonal pricing were highly persuasive sticks. Smart Money was, in Charles McNeer's phrase, "practically all carrot." For its industrial and commercial customers, WEPCO offered interest-free loans and other financing packages for energy-efficient heating, cooling, and ventilation systems, as well as cash rebates for approved motors, drives, lights, and other equipment. Residential customers could take advantage of rebates and loans for refrigerators, freezers, air-conditioners, water heaters, and lighting fixtures, provided the appliances had light appetites for energy. Hoping to rid the region of energy dinosaurs, WEPCO offered $100 U.S bonds for every old refrigerator, freezer, and air conditioner turned in by its customers, and the utility even hauled the appliances away for free!

Response to the program was overwhelming. Some of the system's largest customers completely redesigned their plant and office systems. J.I. Case, the Racine-based manufacturing giant, upgraded its facilities from top to bottom, saving the company's stockholders $100,000 in yearly energy costs. The Milwaukee Public Schools used a $2.7 million interest-free loan and $1.9 million in rebates to cut the system's annual energy consumption by nearly 17 million kilowatt-hours. In Smart Money's first six months, residential customers purchased 65,000 new appliances and turned in 45,000 old ones. Inquiries about the program peaked at more than 1,000 a day, and the company's pick-up crews hauled away 2,000 appliances every week.

Wisconsin Electric had a runaway success on its hands, and the Public Service Commission gladly extended the program. Both the company and the PSC viewed Smart Money as an investment; every dollar spent on conservation was, after all, one less dollar that had to be spent on new power plants. The program's costs were added to the rate base, assuring a return on the investment, and the return increased as the company met specific targets. In 1988 Smart Money's long-term goal was raised from 130 to 300 megawatts of "avoided capacity" by 1993. With more than 10,000 conservation projects under way in the region's businesses and more than 100,000 appliances turned in by mid-1989, no one considered the challenge insurmountable. Smart Money passed the 250-megawatt milestone at the end of 1990, and the program's torrid pace continued.

The system's judicious balance of supply and demand was maintained after 1985. Despite constant flux on the supply side, WEPCO's total capacity grew at an annual rate of 1 percent through 1990. With a major boost from the scorching summer of 1988, peak demand rose considerably faster, increasing 5 percent annually in the same five years. Summertime pressures prompted the company to plan peaking plants for the mid-1990s, but supply and demand conformed to the same general curve. Despite softer earnings in the late 1980s, electric rates continued to moderate, while the company's stock nearly doubled in price and its public approval rating soared to 97 percent. "Overall," said Charles McNeer in 1989, "our results have been as impressive as any in our industry." His peers agreed emphatically. Wisconsin Electric won the Edison Award for 1990, the highest honor in the American utility field.

Smart Money Energy Program

Smart Money was the most ambitious of the demand-side efforts. Its centerpiece was an appliance turn-in program that ultimately retired more than 350,000 energy hogs.

A Civic Powerhouse

The major story of the 1975-1990 period was undoubtedly the well-crafted interplay between supply, demand, and the balance sheet. It was not, however, the only story. Just as Charles McNeer transformed the company's operational side, shifting the dynamics of supply and demand into low gear, he also transformed WEPCO's presence in the public arena. What McNeer wanted was a fundamentally new relationship with the world around the company. The result was the rebirth of a quiet, even sleepy, utility as a civic powerhouse.

Since the end of World War II, Wisconsin Electric's executives had been content to play an essential, but essentially anonymous, role in the community. Electricity was the reassuring hum in the background of daily life, and the company that produced it had no interest in raising the volume. The multiple dislocations of the 1960s and '70s forced Wisconsin Electric to change its tune. As environmentalists, consumer advocates, and politicians claimed larger roles in the utility's affairs, dignified silence was no longer an option. In 1974 John Quale said, "When a small vocal cadre seeks to identify business as an enemy of 'The People,' business must respond." But Quale's response was basically passive. WEPCO answered questions from the media, sent representatives to hearings, and replied, sometimes vehemently, to the charges of its critics, but the company rarely moved beyond defensive communication.

Charles McNeer sought a more assertive and more constructive public presence for his company. In October, 1975, only a few weeks after assuming the presidency, McNeer informed his management staff that change was on the way:

> *No longer is it enough to go about quietly doing our job. Our customers have a right to know what we are doing and why. Our stockholders have plenty of places to invest their savings, and will entrust it to us only if they know how we operate and why we make the decisions we do. Public officials must know they can get credible and useful information from us on energy matters as they try to make responsible legislative and administrative decisions....*
>
> *We will depend much less on newspaper headlines and much more on person-to-person and person-to-group communications. We will shed both our total dependence on the mass media and our undeserved identity as the cold, impersonal corporation in communicating with customers, community leaders, employees and investors.*

Charles McNeer fielded questions from the audience at a 1979 stockholders' informational meeting.

In 1976 McNeer hired John Speaker, one of the most aptly named communications executives in the business, to head his new campaign. An Oklahoma native and a Texas utility veteran, Speaker launched a number of programs, including, of course, a speakers' bureau. Practically no constituency was overlooked. McNeer and his fellow officers took to the road, convening stockholder meetings in Appleton, Kenosha, and Iron Mountain. The same group met regularly and informally with local business leaders and media representatives. The company opened hotlines for stockholders, contractors, and reporters, and the Communications Department took on a full complement of writers, editors, photographers, meeting planners, lobbyists, and media relations specialists.

The communications effort extended, significantly, to the company's natural enemies. The tide of anti-utility protest had continued to rise after 1975. Controversies over air pollution, nuclear waste, and higher rates still raised tempers, and there was a dependable crop of new issues, including the effects of stray voltage on the region's dairy herds and the impact of electromagnetic fields on public health. New advocacy groups emerged to keep the various issues in the public eye. The most important was the Citizens' Utility Board, a voluntary membership group established by the state legislature in 1979. CUB became Wisconsin Electric's most persistent gadfly, filing briefs and issuing press releases designed, in the view of the group's leaders, to make the company more accountable to its customers. The Utility Board's mandate covered all the state's utilities, and the legislature gave the group a highly unusual recruiting tool: Wisconsin utilities were obliged to insert CUB's appeals for members and donations in at least four of their monthly bills.

Charles McNeer's response to criticism, whether it came from CUB or the state's older environmental groups, was unfailingly temperate. "We will not," he said in 1975, "communicate by confrontation." In 1977 McNeer began to host "WE Listen" luncheons for the leaders of groups representing consumers, women, minorities, and sundry other political interests. Three years later, the listening effort was formalized in the Consumer Advisory Council. Composed of labor leaders, minorities, farmers, environmentalists, and consumer activists — among them some veteran utility-bashers — the Council was anything but a sympathetic sounding board. Some WEPCO veterans accused McNeer of fraternizing with the enemy, but forming the Council sent a message: Wisconsin Electric will not be embattled. Communication replaced confrontation, and the siege mentality of the 1960s became a fading memory.

The Advisory Council was convened by WEPCO's Consumer Affairs Department, which became a formal part of the corporate structure in 1981. The new department's head was Nancy Noeske, a former public school teacher (with a doctorate in education) and a longtime community activist. Noeske's department took charge of everything from plant tours to community development work, and her staff came to include other activists whose social and political perspectives differed radically from those of the average engineer. Promoted to a vice-presidency in 1986, Noeske became the first woman executive officer in the system's history.

CUB

CITIZENS' UTILITY BOARD

CUB was the most tenacious of the watchdog groups protecting what its leaders considered the public interest.

The Consumer Advisory Council, convened in 1980, opened the company to new viewpoints. Some were distinctly hostile.

Discussing the issues with outside "advisors" helped to clarify opposing points of view. When those issues reached the legislative arena, however, McNeer's approach was a good deal more direct than luncheons and advisory councils. The impact of public policy on WEPCO's affairs was absolutely decisive, and McNeer took pains to ensure that his company's voice was heard at the decision-making level. His goal was to broaden and amplify Wisconsin Electric's input. Although the utility had been lobbying diligently since the days of Henry Payne, its leaders had long enforced a systematic ban on direct political action. In 1956 the company had issued an edict forbidding its employees to run for city, county, or state office. Even candidates for suburban school boards were required to obtain permission from their supervisors.

McNeer lifted every ban on political activism. Although he had no interest in adversarial relationships and even less in partisan politics, he was determined to get the company's message through to legislators and regulators. WEPCO's lobbying efforts (which McNeer preferred to think of as "education") were expanded steadily. A political action committee was formed in 1976 to support Wisconsin candidates "who favor our positions on utility issues." It was followed by a federal PAC in 1978 and a PAC for Michigan candidates in 1988. Employees were urged to participate in the "Self-Government and You" program, a series of one-day civics classes first offered in 1982. Enrollment reached 500 within six months, and by 1984 at least eight WEPCO workers had won election to local offices. In 1986 the company began to sponsor "Good Citizenship" seminars, half-day programs that dealt exclusively with the nuts and bolts of political campaigns.

Political activism was obviously in the system's self-interest, but McNeer also led Wisconsin Electric into the broader waters of volunteerism. He led, characteristically, by example. Displaying a boundless appetite for involvement, WEPCO's chairman seemed to be everywhere in the 1980s. McNeer served on the boards of more than a dozen non-profit groups, ranging from The Nature Conservancy to the Association of Commerce, and he rose to the chairmanship of several. Whether the matter at hand was a sports arena for downtown Milwaukee or a scholarship program for inner-city students, Charlie McNeer figured prominently in the discussions.

After years in the shadows, his company followed suit. Disturbed by proposals for "lifeline" rates intended to help the region's poorest families, WEPCO decided to act on its own. In 1983 the utility established the Good Neighbor Energy Fund to provide energy grants and counseling to elderly, disabled, and low-income residents of its service area. Launched with a

A Great Man In a Great Place

Civic Leader Charles McNeer Has Spurred Downtown Development and More

Charles McNeer

As winner of the Press Club's 1986 Headliner Award, Charles S. McNeer is recognized by the Press Club for his significant community leadership and numerous civic contributions.

Chairman of the Board and Chief Executive Officer of Wisconsin Electric Power Co., McNeer's most public efforts have centered on economic development and downtown revitalization.

In 1982, McNeer and Wisconsin Electric recognized the limited scope of state efforts at economic development and launched an aggressive and wide-ranging economic development program. The program has created or retained more than 3,000 jobs since its start.

Milwaukee itself will long bear the marks of McNeer's civic leadership. A key player in downtown revitalization, McNeer, president of the Greater Milwaukee Committee, has played a major role in site selection for the Bradley Center, and was instrumental in donating Wisconsin Electric's East Wells Power Plant to the Milwaukee Repertory Theater as the keystone of the Theater District redevelopment project.

McNeer serves on the boards of the Metropolitan Milwaukee Association of Commerce, Forward Wisconsin, Milwaukee Redevelopment Corp., Milwaukee Innovation Center, Competitive Wisconsin, the Federal Reserve Bank of Chicago, the Nature Conservancy, Family Service America and Wisconsin Electric Utilities Research Foundation. He chairs the executive board of the YMCA, the Wisconsin Utilities Association and the American Committee on Radwaste Disposal.

McNeer's civic activism won him numerous honors, including the Milwaukee Press Club's Headliner Award for 1986.

$100,000 gift from the company and swelled by donations from employees and customers, the Fund's assets climbed to $1 million in 1987. Wisconsin Electric branched out into the community development field in 1986, offering grants and loans to neighborhood groups who were rehabilitating blighted properties. By 1990 the company had formed "development partnerships" with 215 community organizations. Other activities ranged from support for Wisconsin colleges to the outright donation of the East Wells powerhouse to the Milwaukee Repertory Theater.

The company's most notable civic activity was undoubtedly its economic development campaign. WEPCO's interest in the economic well-being of its service area had deep roots. In 1966, for instance, the utility had placed an ad in *Fortune* touting the "Main Street to Metropolis" diversity of its service area. ("No matter what your line," the ad promised, "along our lines there's a community ideal for your plant.") In 1969 the company hired "area development coordinators" to promote "the growth of industry and business" in Wisconsin and Upper Michigan. WEPCO's promotional efforts took on a new urgency in the early 1980s. A sharp recession ravaged the region's manufacturing base, destroying tens of thousands of jobs and generating the highest unemployment rates since the Depression. Wisconsin Electric could hardly relocate its operations; any threat to the service area's economy was viewed as a direct threat to the company. At the

1982 annual meeting, McNeer declared it vital to "restore the area's economic health." He pledged WEPCO's best efforts "to maintain the area's strong industrial base by retaining the industries and jobs now located in our service territory, by helping those existing industries to grow or expand in Wisconsin, and by attracting new businesses and industry to Wisconsin and our service area."

The result was an aggressive marketing campaign. In 1982 Patrick Le Sage, WEPCO's new manager of industrial development, described southeastern Wisconsin as "one of the most underpromoted areas in one of the most underpromoted states in the United States." Determined to close the awareness gap, the company filled national and regional business magazines with ads on a variety of themes: "Wisconsin Business Is In A Fine State," "The I-94 Connection," and "Wisconsin: Where Success Stories Start." Perhaps the most novel promotion was a billboard placed at the Illinois border in 1982. Targeted to business executives heading home from Wisconsin vacations, it asked simply, "Why Go Back?," and offered a fresh alternative: "Move your business to Wisconsin instead!"

Wisconsin Electric's development efforts were not limited to catchy slogans and clever signs. The company's missionaries worked directly with unbelievers in neighboring states, tirelessly preaching the gospel of Wisconsin. On one occasion, the utility offered its own central stores facility to a firm that was looking for a plant in the region. As inquiries multiplied, the company considered developing the land adjacent to its power plants as industrial parks. All these activities led directly to the formation of the Wisconsin Energy Corporation, a venture described in the next section, but they had a positive impact on the region's economy years before the utility grew its own holding company.

Tweaking their counterparts in Illinois, Wisconsin Electric's economic development specialists placed a provocative billboard at the state line in 1982.

In areas as diverse as industrial development, philanthropy, and political activism, Charles McNeer forever altered his company's role in the world around it. He changed Wisconsin Electric from a spectator to a player, from a defensive giant to a civic powerhouse. The change was more than a matter of emphasis. Blending communication with activism, McNeer succeeded where his predecessors had not even tried: in building a common-fate mentality among the company's various publics. Consensus of any kind had been a scarce commodity before 1975. In the polarized atmosphere of the 1960s and early 1970s, environmentalists had been dismissed as irresponsible tree-huggers and utilities had been derided as rapacious profit-seekers. What had been overlooked was the fact that America rested on a technological base that was powered, in large part, by electricity. There was ample room for honest disagreement after 1975, but McNeer succeeded in communicating a message of the most basic importance: Electricity is all of us.

The Rational Workplace

Whether he was courting industrial prospects or promoting conservation, Charles McNeer could depend on only one tool to bring his plans to life: the Wisconsin Electric work force. It was the system's people who made the multiple programs work, not board decisions or executive edicts. McNeer, therefore, spent as much time on the company's internal development as he did on issues of supply and demand. The result was a transformation of the workplace every bit as sweeping as the revolution in the company's relationship with the outside world.

The organization that McNeer inherited was a blend of the modern and the traditional. Slowly but inevitably, Wisconsin Electric had become more like other American businesses of the late twentieth century. In 1972, for instance, most of the health care and life insurance plans offered by the Employes' Mutual Benefit Association were transferred to private carriers like Blue Cross and Prudential. The in-house medical staff was reduced to a skeleton crew, and its work was limited to employment physicals and treatment of minor injuries. The EMBA's officers concluded that the rest of the world had finally caught up with benefits the Association had been offering since 1912. Wisconsin Electric introduced a variety of other thoroughly modern programs, including a stock ownership plan in 1977, a substance abuse intervention program in 1979, and flexible work schedules in the same year. (Nearly 80 percent of

Employee initiatives of the McNeer years included a greater emphasis on workplace diversity (above) and a new commitment to internal communication.

the employees on "flextime" chose to begin their days earlier than usual.) The company also placed greater emphasis on workplace diversity. WEPCO hired its first affirmative action officer in 1980, and the ranks of women and minorities grew steadily.

Progressive programs went hand in hand with an aggressive communications effort. Determined to create an informed work force, John Speaker's staff turned out *Telenews* daily, *Currently* biweekly, *Outlet* monthly, *Synergy* quarterly, and *Televiews* periodically. Regular surveys of employee attitudes began in 1977, and management tried to address the most frequently expressed concerns. When workers outside Milwaukee complained of their isolation, officers went out to meet them. When others griped about opportunities for advancement, the company began to post its openings for management jobs. New channels of communication cut across departmental lines; confidential formats like the Write to Know program and a variety of face-to-face

forums provided ample opportunities for input. Wisconsin Electric was a sprawling kingdom, employing 6,000 people and covering an area of 12,600 square miles; Charles McNeer wanted to ensure that it was a kingdom of common knowledge.

Although it was, in many respects, an up-to-date, responsive employer, WEPCO was also a stronghold of tradition. The system's employees participated in activities that their parents and even grandparents would have found familiar, from bowling leagues and sheepshead tournaments to singing groups and Sunshine Clubs. The EMBA had long since shed its aura as "a spirit-forming and spirit-developing society," but the Association remained the umbrella for WEPCO's social and recreational programs. Basketball and softball teams with names like Coal-Blooded and Nuclear Waste competed in company leagues, and a new travel club carried members as far away as Switzerland and Hawaii. But the most durable company tradition was the longevity of its employees. In the late 1970s, more than a third of the system's workers were veterans with at least 20 years of service. In a 1982 survey, employees were asked what they liked most about their jobs. Fifty-six percent ranked job security first, more than twice the number who placed wages and benefits at the top. Conditions would change utterly in the years just ahead, but Wisconsin Electric was a company whose workers came with every expectation of staying.

Whether they were veterans or relative newcomers, traditional or modern, Charles McNeer had high regard for his thousands of co-workers. "We have outstanding people at all levels of management," he said in 1975. "We have capable and dedicated employees." But McNeer also saw room for improvement. His central goal was greater rationality in the workplace. He wanted the company's employees to develop a systematic understanding of what they did and why, and he sought structural changes designed to fortify the company's internal logic. McNeer was not as interested in making people work harder or smarter as he was in making the entire organization work *better.*

EMBA softball leagues continued a tradition dating back to 1913. This team won the company championship in 1987.

The campaign for a rational workplace produced both structural change and physical change. McNeer began to ponder structural questions almost as soon as he took office. In early 1976, he realigned the company's departments according to function — Division Operations, Engineering and Construction, Customer Relations, and others — and put each group under a single executive officer. A few months later, Wisconsin Natural Gas took over the gas operations of Wisconsin Michigan Power, a move that simplified negotiations with the pipeline firm that supplied both subsidiaries. On January 1, 1978, McNeer made a significantly bolder move: Wisconsin Electric absorbed Wisconsin Michigan Power entirely. Most "Wis Mich" veterans resented what they perceived as a loss of independence, but McNeer was convinced that the merger would "reduce administrative and regulatory duplication and present opportunities for further efficiency and effectiveness in overall operations." The former subsidiary became a division, and WEPCO was a two-company system again for the first time since the early 1900s.

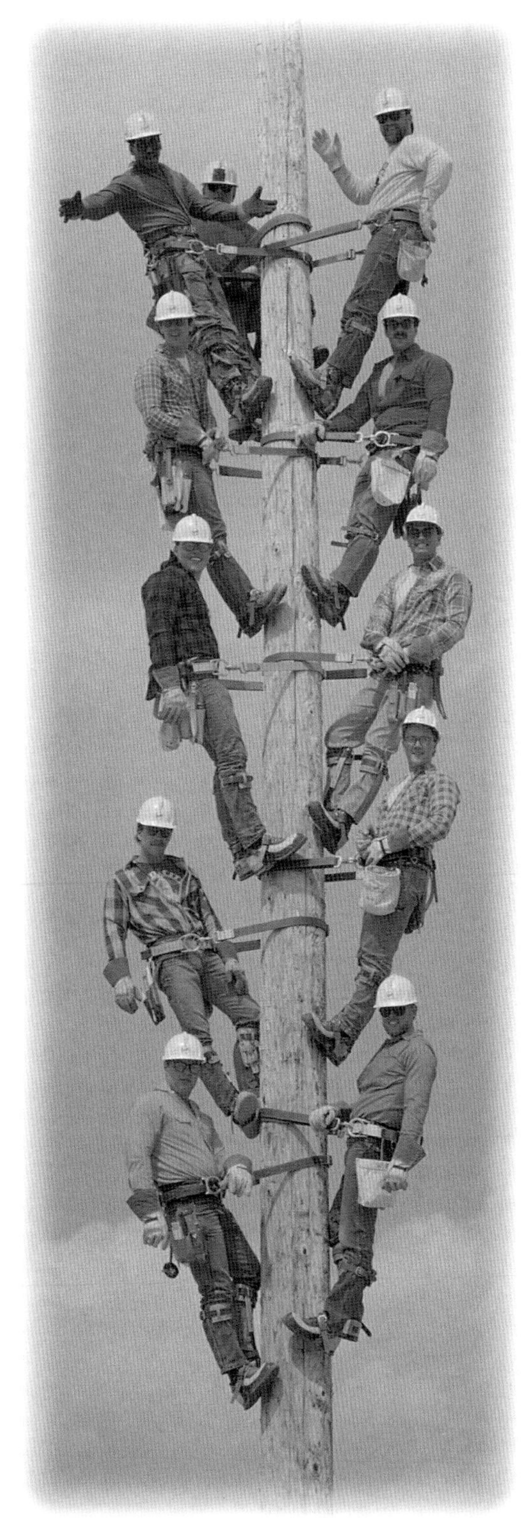

Charles McNeer introduced several programs designed to take employees on every level to new heights — including the line mechanics' class of 1990.

As he rationalized the system's internal structure, McNeer also tried to transform the daily routines of its 6,000 employees. Management tools go in and out of fashion constantly in corporate America, and Wisconsin Electric tried most of them, including team-building, quality circles, and even, for a time, transactional analysis. But the tool with the greatest impact was management by objectives. In January, 1979, after three years of discussion, McNeer announced an MBO program for his employees. Every work group was asked to prepare a detailed "responsibility statement" and a set of quantifiable goals for meeting its objectives. The results, McNeer hoped, would be a great leap forward in employee creativity and improved productivity for the system as a whole. "Advancing technology will offer some efficiencies," he said, "but the most important advances in our effectiveness and efficiency must come from our own ingenuity."

Management by objectives made employees take a more intentional approach to their jobs, but McNeer had limited success in changing actual behavior. In 1979 he promoted MBO as a way to "increase employee participation in management by delegating significant responsibilities further down in the organization structure." Ten years later, he was still talking about the need for "a more participative style of management." McNeer found his initiatives ambushed by a corporate culture that was nearly as old as the company. For all its progressive policies, Wisconsin Electric was a bureaucracy at heart, a massive enterprise with nearly 1,200 job titles, formidable walls between departments, and a management structure that had more layers than a Viennese torte. The system had enough inertia to swallow practically any attempt at reform. It was not until the 1990s, when competitive pressures forced a change in thinking, that issues of culture were finally resolved.

McNeer had better luck solving a space problem in ways that embodied his desire for a rational workplace. Since its dedication in 1906, the Public Service Building had been the Vatican of the Wisconsin Electric system, the single point at which all lines of authority converged. A gradual loosening of those lines was evident in the years just before McNeer took office. In 1968, for instance, WEPCO announced the formation of its Waukesha Division as "a first step in a decentralized system of operations." Three years later, the company broke ground for a central stores facility in the Town of Pewaukee, nearly 20 miles west of downtown Milwaukee. The facility was completed in 1974, and it was joined in 1975 by the System Control Center, which moved, computers and all, from the basement of the Public Service Building. Although the company's nerve center moved west, the PSB remained its administrative center, and the Michigan Street landmark was soon bursting at the seams. With the continuing growth of downtown employment, nearly 1,800 workers were scattered among five buildings, a situation that made absolutely no organizational sense.

Charles McNeer took steps to *re*centralize WEPCO's administrative work force. In 1977 the company unveiled plans for three separate projects: a 12-story headquarters building, a

WEPCO's attempts to decentralize were obvious in 1974, when the System Control Center began to take shape in suburban Pewaukee.

combined substation and parking garage, and a cogeneration plant with a capacity of at least 250 megawatts. Plans for the power plant, which would have stood just west of the Summerfest grounds, were quietly dropped in 1979; WEPCO had little need for its capacity, and there was entrenched resistance to downtown smokestacks. Planning for the headquarters building continued, but the details changed constantly. Architects submitted plans for buildings of 13 stories, 26 stories, and 15 stories, with 19 emerging as the "final" choice. The building's site was never in doubt. Between 1968 and 1972, Wisconsin Electric had purchased the two blocks south and southwest of the PSB — the site of the old Milwaukee Road depot — and was using the land for parking until it was needed for expansion. Surprisingly, given McNeer's personal interest in history, the Public Service Building itself was considered expendable. John Beggs had designed it to support 12 stories, but company officials had no interest in building from the base of 1906. They considered selling or, worse yet, demolishing the landmark as soon as the new headquarters was completed.

A move to recentralize was afoot in 1982, when the Public Service Building's board room was restored to its original grandeur.

In 1980, a year of record-high interest rates, WEPCO put its building plans on hold. "The present economic facts of life," McNeer explained, "dictate that we reduce or delay spending where we can." McNeer and his colleagues, in the meantime, decided that the Public Service Building was not disposable after all. In 1982 the company restored the old board room to its original elegance, and in 1983, when interest rates had dropped to more reasonable levels, WEPCO announced plans for a five-story, $49 million "annex" linked to the PSB by a skywalk. Work on the adjoining parking garage/substation was scheduled first. Completed in 1984, it won instant favor among employees as a site for viewing Milwaukee's Great Circus Parade. Ground for the PSB Annex was broken in October, 1984, and the building was completed in December, 1986. With its soft earth tones and strong horizontal lines, the final product owed more to the Prairie Style practiced by Frank Lloyd Wright than to the neoclassical models that inspired John Beggs' monument. Behind the elegant lobby were workrooms as big as ballrooms and just as open; the building was

designed with flexibility in mind. The Annex was also, naturally, a model of energy efficiency, with solar water heating, sophisticated heat recovery systems, and a foundation that included copious amounts of fly ash from the Pleasant Prairie Power Plant.

Wisconsin Electric's new Annex was the rational workplace embodied, and its completion was a milestone in Charles McNeer's term as chief executive. It was not, however, the main event in the company's internal development. That distinction belongs to the Wisconsin Energy Corporation. Wisconsin Energy was, in essence, the melding of the rational workplace and the civic powerhouse; it blended McNeer's desires for internal logic and external activism. The holding company's impact on WEPCO's structure was clearly profound. Wisconsin Electric, the perennial parent, became a subsidiary of its own child, and the new enterprise carried the system in directions that McNeer's predecessors could hardly have imagined.

With the opening of the PSB Annex in 1986, Wisconsin Electric had a new focal point. The building's sleek, modern lines served as a counterpoint to the classical splendor of the original headquarters.

It was not an easy birth. Wisconsin Energy's long labor began in June, 1981, when the WEPCO board approved plans for a holding company to "diversify into non-utility operations." The directors wanted to explore "opportunities for growth," and their options included "coal exploration and mining, gas and oil exploration and production, fuel transportation, environmental and laboratory services, cable television, home information services and real estate." (Only the last item became a Wisconsin Energy activity.) In 1980 the Wisconsin Gas Company had established WICOR, its own holding company, simply by notifying the Public Service Commission and transferring its service agreements to the "new" gas subsidiary. Wisconsin Electric fully expected similar treatment, but the PSC, under substantial legislative pressure, decided that it had the power to withhold permission. When federal securities officials and Wisconsin's attorney general agreed, Wisconsin Energy was apparently stillborn.

New developments gave the case a new urgency. In 1982 a severe recession was ripping through the state's economy like a tornado, and Wisconsin Electric's development staff was doing all it could to control the damage. At the same time, the utility was beginning to reap the rewards of its conservation program. Pleasant Prairie's second unit was the last item on the construction agenda, and WEPCO looked ahead to a time when it would have an abundance of cash to invest. What better place to invest it than Wisconsin? What better way to stimulate the state's economy and ensure the utility's ultimate prosperity? In 1983 the company refiled its application to form Wisconsin Energy.

Stalled on the regulatory level, WEPCO turned to the legislature itself for permission. Some critics there charged that the company was simply trying to escape the nets of regulation, and others contended that Wisconsin Electric was "hoarding" the growing pile of cash in its depreciation reserve. The company did, in fact, want to enter fields in which the market, not the government, determined rates of return, but its interest in economic development was equally obvious. In 1985, after two years of legislative near-misses, Charles McNeer stated the case for Wisconsin Energy as clearly as he could:

> *Economic development is not a partisan issue. It's the number one challenge facing the state today. [The holding company bill] will allow us to put our investor resources to work for Wisconsin's economic development. We feel this is an issue where everyone wins.*

After a long and difficult labor, Wisconsin Energy was born in 1987. The new child instantly became the parent of the WEPCO system.

As the recession continued to chip away at the state's economic foundation, economic development became a sort of Holy Grail for public officials. Accompanied by estimates of a $30 million investment and 35,000 new jobs over a 10-year period, McNeer's case was compelling. The Wisconsin legislature passed the bill in October, 1985, and Gov. Anthony Earl quickly signed it into law. Wisconsin Energy came into the world, alive and well, on January 1, 1987, and Wisconsin Electric, for the first time since the break-up of the North American Company in 1947, became a subsidiary again.

Wisconsin Electric and Wisconsin Natural Gas immediately took places under the infant's wing, but most of the attention went to their three new siblings. Wisconsin Energy came into being with three major non-utility subsidiaries: Witech provides venture capital for new businesses (with an emphasis on high technology), Wisvest supports existing businesses (including those under threat of hostile takeover), and Wispark engages in real estate development. Of the three, Wispark quickly assumed the highest profile. In 1987 the company broke ground for LakeView Corporate Park, a 1,200-acre development adjacent to the Pleasant Prairie Power Plant. With excellent access to both Milwaukee and Chicago via Interstate 94, LakeView developed a backlog of prospects almost immediately. Its early "residents" ranged from an upscale outlet mall to a Rust-Oleum plant. Two years after opening LakeView, Wispark launched another project in downtown Racine: Gaslight Pointe, a luxury residential complex on a spectacular lakefront site. Business at both developments was brisk, and Wispark quickly broadened its horizons.

Wisconsin Energy's first annual report provided a simple rationale for diversification:

> *The economic health of our companies depends on the economic health of our customers. Therefore we are investing in new businesses, new technologies, and new ideas for our service area. We are working to preserve jobs and create new ones.*

One of the system's most visible non-utility ventures was Gaslight Pointe, a luxury residential development on the site of the old Racine gas works.

A new generation of senior executives rose to the top in the late 1980s. Bill Boston (upper) and David Porter were two of the most prominent.

The importance of the new structure should not be overstated. WEPCO and Wisconsin Natural had a new parent, but there was no doubt that the utilities were the dog and the other subsidiaries were the tail. In 1987, for instance, non-utility ventures represented precisely 0.86 percent of Wisconsin Energy's assets; electricity and gas accounted for all the rest. But diversification gave the enterprise a flexibility it had never had, even in the free-wheeling days of John I. Beggs. Wisconsin Energy's promise was opportunity, for both the company and the region.

With his latest creation growing nicely, Charles McNeer turned to his last major responsibility as chief executive: management succession. The leaders of McNeer's generation were nearing retirement as a group, and new faces began to appear in the executive suites. Bill Boston, an engineer with broad executive experience at a New York utility, was hired in 1982 as Sol Burstein's understudy. When Burstein retired in 1987, Boston succeeded him as head of the system's power plants and rose quickly to become chief operating officer of the WEPCO subsidiary. David Porter, who had joined the company in 1969 as a nuclear engineer, was named a senior vice-president in 1989, with responsibility for all customer-related activities. But the brass ring went to Richard Abdoo. Raised in the Lebanese community of Port Huron, Michigan, and trained in both engineering and economics, Abdoo had come to Wisconsin Electric in 1975 as director of planning. Six years later, after hearing McNeer praise his "outstanding performance on many key assignments," the board made him a vice-president. Abdoo moved up the ladder quickly, taking charge of one, then three, key departments as a senior vice-president. In 1989 he became WEPCO's president and McNeer's obvious heir apparent. Bill Boston moved into place as second in command, succeeding Russell Britt. More outgoing than McNeer and less blunt than Britt, the system's new leaders represented a significant shift in management style.

After a year of transition, Charles McNeer stepped down as Wisconsin Energy's chairman and chief executive officer on April 30, 1991, precisely one month after Russell Britt's retirement. In a 1989 report to employees on succession planning, McNeer had said, "When Russ and I leave, if we're successful, you won't even notice it." They did notice, of course, and they continue to notice, not because McNeer and Britt were unsuccessful but because they were so successful. There is little doubt that Charles McNeer's administration will be remembered as one of the most eventful of the company's first century. His 15 years at the helm were filled with tangible achievements: Wisconsin Energy, a state-of-the-art office building, a handful of new power plants, a much larger assortment of highly effective conservation programs, and absolutely stellar financial performance. Every $100 invested in Wisconsin Electric stock in 1980 was worth more than $800 in 1990.

By the time of his retirement in 1991, Charles McNeer, ably backed by Russ Britt, had made Wisconsin Electric a model, once again, for other utilities to emulate.

His tangible accomplishments were self-evident, but McNeer's most enduring contribution was intangible. With substantial help, he made Wisconsin Electric a different *kind* of company between 1975 and 1990. During the long postwar boom, running the utility had not, in truth, been an overwhelming challenge; growth was practically automatic. As the old order vanished in the decade after 1965, WEPCO adopted an essentially reactive stance, draining its energies in pursuit of business as usual. McNeer moved the company's posture from the reactive to the intentional, from driven to driving. On virtually every issue, from conservation to communication, he broke from the pack in pursuit of a future that made sense for his company. McNeer found that future. It is the position of strength from which his successors operate today.

Chapter 10

1990-1996

Reengineering the Enterprise

In 1990, shortly after he was named Wisconsin Electric's chief executive officer, Richard Abdoo talked about the expectations he faced in his new job. "What do I do for an encore," he asked rhetorically, "after what Mr. McNeer has done?" Abdoo's performance has amounted to much more than an "encore." As soon as he took center stage, WEPCO began perhaps the most dramatic transformation in its history. Not since 1952, when the transit system was spun off (and with it half the corporate work force), has there been such pervasive change in the company's operations. Not since the World War I era, when James Mortimer shifted its focus from traction to power, has there been such a fundamental rethinking of the utility's place in the world. WEPCO has asked radical questions and found equally radical answers, answers that have led, ultimately, to the threshold of a merger with one of its largest neighbors. The reengineering process has been trying, even traumatic at times, but WEPCO's leaders believe that nothing less than a new incarnation will ensure the system's continued prosperity.

It has not all happened at once. Dick Abdoo's starting point was his conviction that change in the outside world made it absolutely necessary for his company to change. An old order vanished, in Abdoo's view, during the economic dislocations of the late 1970s and early 1980s, when the global marketplace emerged as a key factor in business planning. Companies intent on prospering in the new order, particularly those shaken by the 1981-82 recession, tried to slim down and speed up to keep pace with their competitors around the world. They did so, in part, by forging new relationships with their suppliers. Any company hoping to do business with an automaker or a steel firm, for instance, had to meet entirely new standards of quality, speed, and economy.

Any company, that is, except the local public utility. When industries in the throes of restructuring tried to forge new relationships with their energy suppliers, they encountered a series of brick walls. As "regulated monopolies," America's electric and gas utilities did not live by the rules of the marketplace. They operated on a cost-plus basis; their rates reflected the cost of providing energy plus something extra for the shareholders. But the very idea of monopoly, "natural" or otherwise, seemed increasingly anachronistic in the late twentieth century. Deregulation, or at least reregulation, was advanced as a means to bring competition, and thereby lower costs, to the utility industry. Slowly, by degrees, the climate began to change. There were no sudden moves on the part of either regulators or customers, but utility executives began to look ahead to changes on the

Richard Abdoo, chief architect of a future that will differ radically from Wisconsin Electric's past

same scale as those sweeping the telecommunications, airline, and interstate banking industries. "It's not a matter of 'if,'" said Abdoo repeatedly. "It's a matter of 'when.'"

First Steps

There were faint stirrings of the new competitive order in the last years of Charles McNeer's presidency. Since the bitter turf wars of the 1920s, the borders between America's electric utilities had been practically chiseled in granite; customer shopping was discouraged, and "poaching" was practically unheard of. The lines blurred noticeably in the 1980s. On May 1, 1986, after protracted discussions, Wisconsin Electric began to provide power to Geneva, Illinois, a Chicago-area community that had been one of Commonwealth Edison's wholesale customers. A home-state competitor returned the favor a year later; in 1987 WEPCO locked horns with Wisconsin Power & Light over service to an industrial park in Deerfield (near Madison) and to the municipal utility in Elkhorn. The outcome was a draw: Wisconsin Electric lost in Elkhorn but won in Deerfield.

The border skirmishes of the 1980s served notice that the walls were beginning to come down. Charles McNeer sounded a warning at the 1987 spring management conference:

> *Utilities, if they are to remain successful, are going to have to become competitive and more innovative.... Active competition for municipal and industrial loads can be expected to come from neighboring utilities. More and more utilities are moving across service boundaries seeking new customers — and some customers are shopping around.*

Wisconsin Electric's rates were (and are) among the very lowest in urban America, but they were not the lowest in Wisconsin — a fact that gave the company's leaders cause for pause. At the same spring conference, Russell Britt described competition as "the most important problem" facing WEPCO in the near term. "If we're to meet the challenge of competition," Britt continued, "we've got to look harder at programs and people."

The result was Operation Compete, an ambitious campaign to trim the system's expense base by 10 to 15 percent. Announced in 1987, the effort affected every department's operating budget, but the burden fell most heavily on the work force. Between 1987 and 1989, a combination of early retirements and layoffs reduced system-wide employment — both electric and gas — from 6,125 to 5,610, a drop of more than 9 percent.

Operation Compete was a fairly mechanical response to the specter of increased competition. McNeer and Britt knew that it would take more than budget cuts and layoffs to make Wisconsin Electric and Wisconsin Natural Gas "the lowest-cost regional providers of energy services" — top management's ultimate goal. What the future demanded, in the view of the system's executives, was a new way of thinking, a shift in focus from the bureaucratic to the entrepreneurial. In 1988, as the cost-cutting continued, the company unveiled Best in the Business, an attempt to redirect the energies of the WEPCO work force. More a set of policy objectives than a guide to action, Best in the Business highlighted the need for improved customer service, more aggressive marketing efforts and, in Russ Britt's words, "a work environment that encourages employee participation, supports innovation, recognizes employee contributions and rewards performance." After living under a command-and-control structure for decades, the company was trying to diffuse power, and with it responsibility, throughout the organization.

The effort met with limited success. In 1989, more than a year after announcing Best in the Business, McNeer and Britt were still stressing the need to turn WEPCO "from a cost-plus utility to a customer-oriented competitor," and senior managers were still trying to broaden the circle of responsibility. Affirming his own desire to empower those below him, Britt said, "I'm going to let go, let go, let go." Not everyone shared his taste for delegation. Veteran supervisors, in particular, found it difficult to share hard-won power with their subordinates. In a 1990 survey of employees, 44 percent believed that management wanted to give them more authority to make decisions, but only 7 percent felt that it was actually happening. In the meantime, some of the economies realized during Operation Compete were slipping away. After dropping to its modern low point in 1989, system-wide employment began to creep up again.

The problem, senior managers concluded, was Wisconsin Electric's culture. The company had developed, over the decades, a body of beliefs and behaviors — a culture, in a word — that set it apart from other business organizations. That culture was, above all, conservative. Reflecting WEPCO's status as a regulated monopoly, its emphasis was on security — for customers, for regulators, and not least of all for employees. There was a place for everything, and everything was in its place, all conforming to an elaborate canon of rules. New employees soon learned that certain qualities were valued more highly than others, among them loyalty, perseverance, and a plodding attention to detail. Wisconsin Electric did things thoroughly and well, but it was not a congenial place for those who wanted to do things quickly.

Charles McNeer had demonstrated that the old organization could learn new tricks; shifting the system's focus from consumption to conservation was among his stellar accomplishments. But McNeer had used the old command-and-control mechanisms to meet changed circumstances in the outside world; he achieved dramatic results within the constraints of the existing culture. McNeer's approach was precisely right for the turbulent conditions of the 1970s and '80s, but the rise of competition created turbulence of a different order. Shifts in the marketplace called for a nimbleness, a corporate agility that WEPCO's traditional culture stubbornly resisted. It became clear that a determined manager could

The sheer size of the WEPCO system, symbolized by this giant circuit breaker at Pleasant Prairie, hindered even the most serious attempts at change.

Whether he was addressing active employees or retired veterans, Dick Abdoo stressed the need for a transformation.

command employees to take orders, but not to take individual responsibility for their company's success.

By the time Dick Abdoo began to take the reins in 1989, issues of culture had moved to the foreground. There were undoubtedly managers who found the inherited culture sheltering; Abdoo found it absolutely smothering. "We had this entitlement philosophy that drove me crazy," he recalled. "People came here looking for security and stability, and they found it. It was a cradle-to-grave sort of culture." The disadvantage was that security was valued over creativity. Too many bright newcomers either lost their sparkle or left the company entirely.

The company's new leader unashamedly offered himself as a role model for the culture he hoped to create. Dick Abdoo is the ultimate shirtsleeves manager, a man impatient with protocol and hungry for feedback. He also acknowledges a temperamental restlessness and a definite fondness for ferment, for stirring things up. From the earliest days of his career, Abdoo distinguished himself as a risk-taker — no small achievement in an organization better-known for less assertive qualities. Soon after taking charge, he launched a campaign to remake WEPCO's culture in his own entrepreneurial image.

His first step was relatively modest. In 1989 Abdoo expanded the strategic planning team from a handful of top executives to a group of 57 managers from every corner of the company. All were asked to develop plans for departments outside their own areas of expertise — an attempt, Abdoo told the board, to broaden their thinking and, not incidentally, to test their abilities "to adapt to the new environment in which the companies will be operating."

The new president's next step covered significantly more ground. In 1990, at Abdoo's direction, WEPCO began to offer a training program called DELTA — Developing Excellence through Leadership, Teamwork, and Accountability. At its core was a series of group exercises designed to promote cooperation, improve communication, and heighten the potential of every individual to bring about change. By the end of 1991, DELTA had reached 3,000 employees, a majority of the system's work force. Dozens of office walls sported some of the program's more popular slogans, including "Be Here Now," "End Hidden Agendas," and "Expand Your Comfort Zone."

Dick Abdoo continued to turn up the heat. Addressing a group of managers in 1991, he made it clear that change was not optional:

> *As I look ahead, I see a company where employees have the power to make a difference in the lives of customers, fellow employees, shareholders and communities.... Our customers' needs are changing, and we must be able not only to respond, but to anticipate those needs and provide the answers. To do so, we are going to have to change our corporate culture so that every single person in this company participates in fostering change.*

Abdoo worked overtime to communicate the case for change, addressing small groups of employees as well as all-company conferences. By 1992 "changing the corporate culture" had become the operative buzzword at Wisconsin Electric, and Abdoo's "vision" for the future — "You Have The Power To Make A Difference" — was inscribed in brass letters at WEPCO's downtown headquarters.

The various exercises helped to encourage fresh thinking, but stirring speeches and high-sounding slogans were not enough to create a sense of urgency in the work force. In the February, 1992, issue of *Synergy*, the employee magazine, Abdoo went public with some of his frustrations:

> *The difficulty lies in motivating people. Each of us must perform and earn our keep. It isn't good enough that we did a great job yesterday. We have to do it again and again.... Somewhere we got into a belief of entitlement that says I'm owed something because I've been a good employee, I come to work every day, I keep my nose clean and have devoted a lot to the company. Those things don't entitle you to anything.*

Abdoo's comments served notice that the company's old culture was an endangered species, but it was not until the following spring that the future began to take shape. On March 26, 1993, Dick Abdoo announced plans to "revitalize" the Wisconsin Electric system:

> *Our goal is to make certain that our rates are not only lower, but substantially lower, than all our neighboring utilities by 1997.... This will mean major changes for our company and will require the assistance of all employees.... We will have to make major changes if we are going to stay strong in the future.*

All the exercises of the previous few years — strategic plans and pep talks, DELTA sessions and employee conferences — had been nothing more than psychological spadework, attempts to create receptive soil for the "major changes" promised by Abdoo. Within a few weeks of the March 26 announcement, he had hired a consultant, assembled a revitalization team, and directed its members to develop "a new plan for the future organization of the company." The fruit of their efforts was nothing less than revolutionary.

Keeping the Lights On

By the summer of 1993, the workers of WEPCO were operating in two different time zones. The shape of the future — particularly the future of their jobs — was a constant concern, but the demands of the present were every bit as compelling. The service area's economy had not gone into hibernation while Wisconsin Electric geared up for change, and the demand for energy did not slacken as planning teams explored new directions for the utility. With Dick Abdoo spearheading the revitalization effort, day-to-day responsibility for the system fell to Bill Boston, WEPCO's chief operating officer, and Boston had his hands full.

Following policies established during the McNeer years, Wisconsin Electric's leaders tried to maintain a judicious balance between supply and demand. Economic growth and summer heat put specific pressure on peak demand. Between 1990 and 1995, the system's peak increased at an average annual rate of 3.5 percent, and there was simply not enough existing capacity to keep pace. WEPCO's response was unambiguous. In 1992, for the first time in well over a decade, the company broke ground for a new power plant: a peaking station in the Town of Concord, near Watertown. It was actually a complex of four 75-megawatt

Demand for power continued to grow as the company geared up for change. New combustion-turbine plants at Concord (above) and Paris have helped WEPCO stay ahead of the curve.

combustion turbines, all fired by natural gas. Work on a virtually identical plant began in the Town of Paris, outside Kenosha, in 1993. The Concord station was completed in 1994 and the Paris facility in the next year, adding 600 megawatts (roughly 12 percent) to the system's capacity. Although the peaking plants were designed to operate only 5 percent of the time, their construction costs totaled well over $200 million — nearly enough to build a modern major-league baseball stadium.

The system's older power plants received their share of attention. In late 1990 the Point Beach Nuclear Plant celebrated 20 years of service and 128 billion kilowatt-hours of generation — more than five times the power produced by the entire system in a typical year. Despite some troublesome lapses in operating and maintenance procedures, lapses that cost WEPCO $275,000 in fines in 1992, Point Beach remained one of the most efficient nuclear plants in the world. Its future, however, was threatened by the federal government's inability to find a permanent disposal site for nuclear waste. With the plant's storage pool filling up fast, Point Beach was running out of room for its spent fuel assemblies. The problem was solved, at least temporarily, in 1995, when Wisconsin's Public Service Commission granted permission to store the assemblies in "dry casks" — steel-lined containers with walls more than two feet thick — on an earthquake-proof pad adjoining the plant.

As the Point Beach storage pool neared capacity, spent fuel assemblies were sealed in "dry casks" and hauled from the plant by a custom-built transporter.

Perhaps the biggest news on the supply side after 1990 was the plant that got away. With a major boost from federal laws and state regulations, WEPCO had developed a growing interest in cogeneration — the production of steam to drive electrical turbines and power industrial processes at the same time. In 1991 Repap of Wisconsin, a maker of fine printing papers located just down the Fox River from Appleton, decided to replace its aging boilers with a "cogen" plant. Repap needed a strong partner to build and own the facility. As the Appleton area's energy company, Wisconsin Electric was the obvious choice, but there were other candidates. Independent power producers (IPPs) had entered the market, and WEPCO officials faced sharp competition in their attempt to close a deal with their largest Wisconsin customer. The campaign was successful. In late 1991, the papermaker and the utility signed an agreement to build a 160-megawatt cogen facility on the banks of the Fox. Dick Abdoo expressed "pride and joy" that his company had prevailed in head-to-head competition with an independent.

The victory celebration was short-lived. The Public Service Commission had final authority over all new generation in the state, and in 1993 the PSC decided to open the process to competitive bidding. The move was without precedent, in Wisconsin

In-person energy consultation and free appliance pick-up were two popular features of Smart Money, the program that anchored demand-side efforts until its termination in 1995.

or elsewhere, but the Commission felt that it was necessary, in one member's words, "to drive the costs down to the lowest level possible." The Repap project was one of eleven candidates from around the state. In September, 1993, the PSC announced Wisconsin's next power plant: a 215-megawatt cogen facility in Whitewater. Proposed by LS Power, a Montana-based IPP, the plant would produce steam for the state university in Whitewater and electricity that WEPCO was obligated by law to buy at market rates. Finishing second was scant consolation for Wisconsin Electric. Dick Abdoo and his colleagues were outraged, but a court challenge by Repap failed to overturn the decision. If nothing else, the Repap case offered irrefutable proof that competition was no longer a vague threat looming somewhere beyond the horizon. More than one skeptic concluded that Abdoo's decision to revitalize was right on target.

There were no comparable dramas on the demand side of the equation. Conserving energy had become a way of life for Wisconsin Electric customers, and the 1990s brought more of the same. As in earlier years, Smart Money was the primary vehicle. By the end of 1994, the program had reduced consumption by nearly 2 billion kilowatt-hours — enough to spare WEPCO the expense of building a medium-sized power plant. The greatest reductions, of course, resulted from the largest projects; by 1992, for instance, the City of Milwaukee had installed energy-efficient fixtures in all its traffic signals and streetlights, earning taxpayers $1.2 million in rebates. For residential consumers, Smart Money's highlight was undoubtedly the appliance turn-in program; WEPCO customers ultimately rid themselves, for U.S. bonds, of 377,059 old refrigerators, air conditioners, and other energy dinosaurs. Smart Money in its traditional form expired in 1995, one of a host of changes associated with revitalization. Conservation efforts have continued, but their emphasis has shifted from direct financial incentives to technical assistance.

Environmental programs have also taken new forms in the 1990s. When Congress amended the Clean Air Act in 1990, America's utilities began to operate under new and more stringent rules. The law required the nation's coal plants to reduce their emissions of sulfur dioxide, a key contributor to acid rain, by 50 percent from 1980 levels. Compliance was mandatory, but the act included an unusual "carrot." Each utility was granted a certain number of "emission allowances" based on its discharge of pollutants. Those with the cleanest smoke received the greatest number of allowances, and some companies earned surpluses that could be sold or traded — just like corn, gold, or pork bellies — to other energy firms on the open market. The allowances gave all utilities an incentive to over-comply. For those that were close to compliance, buying credits from others provided an alternative to multi-million-dollar retrofittings. Although some critics derided the allowances as "pollution permits," the system recognized that acid rain knew no borders; the law had broad support from environmental groups who saw it as a means to maximize real-world reductions.

With a large push from strict state laws, Wisconsin Electric was ahead of the game. By shutting down its least-efficient plants and switching to low-sulfur coal in its newer facilities, WEPCO met Wisconsin emission standards that were a good deal tougher than those set in the first phase of Clean Air Act enforcement. The company, therefore, had a surplus of emission allowances. In 1993 Wisconsin Electric became one of the first utilities in the nation to enter the allowance market, selling 10 units (each representing a ton of SO_2) to a Columbus, Ohio, energy company. WEPCO has since remained active in the market, selling from its surplus and trading for allowances that will help its plants meet the more demanding standards scheduled to take effect in 2000.

The Clean Air Act regulated what went up the smokestacks, but Wisconsin Electric spent even more energy dealing with what stayed behind: thousands of tons of coal ash. By 1990 WEPCO was selling nearly half its ash, most for use as a concrete additive, and contractors could hardly get enough of the Pleasant Prairie plant's output. The high-carbon ash from Oak Creek was another story. Hoping to create demand for a virtually unsalable waste product, the system's engineers developed a novel technology. They combined Oak Creek's fly ash with sewage sludge from Milwaukee's treatment plant or paper sludge from a Wisconsin mill, then rolled the combination into pellets and roasted it at 2,000 degrees. The organics burned off, leaving "lightweight aggregate," a rock-like material one-half the weight of crushed stone but just as strong. Work on a plant to produce the aggregate began at Oak Creek in the autumn of 1993 and was finished in the following year. Marketed through a subsidiary called Minergy, the new product has aroused definite interest in the concrete industry.

Other environmental efforts have ranged from the purely local to the truly global. In 1990 WEPCO launched Second Nature, a home-office recycling program that saved nearly 250 tons of

Fly ash from the Oak Creek Power Plant is the key ingredient in a lightweight gravel substitute developed for the concrete industry. The material is shown here emerging from a pelletizer.

paper in its first full year. Some of that paper probably ended up in the company's billing envelopes, which were printed on recycled stock beginning in 1991. Other projects have taken Wisconsin Electric far beyond the borders of its service area. In 1994 the utility agreed to help one of the Czech Republic's dirtiest heating plants make the switch from high-sulfur coal to clean natural gas. Another initiative made WEPCO the part-owner of a sizable piece of rain forest in Belize (the former British Honduras). In partnership with The Nature Conservancy and a handful of other utilities, the company launched a sustainable-harvest program in the forest, helping to preserve a resource that transforms enormous volumes of carbon dioxide to life-supporting oxygen. Both the Czech and the Belize projects underscored the international scope of the air pollution problem, and both may earn emission allowances under the U.S. Clean Air Act. Whether they involved waste paper at home or rain forests abroad, the projects of the 1990s have demonstrated the prominent, and permanent, role of the environment in modern utility planning.

"There Is Certain To Be Some Uncertainty"

In any other period, new power plants and new environmental initiatives would be major news at Wisconsin Electric. In the 1990s, they have been distinctly overshadowed by the

From cleaning up dirty skies in the Czech Republic (left) to saving a rain forest in Belize, WEPCO has taken an increasingly global approach to air-quality problems.

In dozens of face-to-face meetings with employees, Dick Abdoo communicated a simple message: We must change.

company's ongoing revitalization efforts. Those efforts have dominated news reports, internal publications, and cafeteria conversations since 1993, and they continue to occupy center stage during the celebration of WEPCO's centennial in 1996.

Dick Abdoo made it clear from the beginning that this revitalization would be different in kind from the periodic reorganizations of the past. For starters, he assembled a task force that was a study in diversity. Abdoo skipped across traditional lines of age, gender, ethnicity, and rank to ensure that he had the best and brightest on his team. Once the task force members had been selected, Abdoo told them to start from scratch, to "invent" a utility on the basis of customer needs rather than past practice. What he wanted was a genuine reengineering: a thorough redesign of the company's structure, not just its operating details. Abdoo spelled out his goals at the 1993 annual meeting:

> *This project is not about downsizing. It's about rethinking the way we do our work to reduce our rates significantly below our competitors' and improve the quality of our service so that we have a competitive advantage. Neither the team, nor the consulting firm, nor I am approaching this with any preconceived idea of how that should be done. We're starting with a fresh sheet of paper.*

In other conversations, Abdoo compared revitalization to a journey. "I know we have to get to the West Coast," he said more than once. "Whether we end up in Portland or San Francisco is less important than getting there." Abdoo trusted his revitalization team to choose the specific route and final destination.

WEPCO's chief traveler took every opportunity to communicate the need for the journey. In an endless round of employee conferences, Abdoo and company stated the case for revitalization: Competition is changing the world in which we do business, and we must either adapt or face extinction. It is better to change than die, they argued, and it is better to change now than later. Events in the outside world helped to build a sense of urgency. Losing the Repap cogeneration project was a wake-up call for those who believed that the old order was somehow imperishable. So was the U.S. Energy Policy Act of 1992, a thousand-page document that eased every utility's access to the transmission lines of its neighbors. The industry was edging closer and closer to "retail wheeling" — the sale of one utility's energy through the lines of another. And the Public Service Commission of Wisconsin, always on the leading edge of regulatory reform, was giving unmistakable signals that it favored more competition rather than less.

The breakup of AT&T into "Baby Bells" and the subsequent free-for-all in long-distance services was nearly always offered as a preview of the brave new world awaiting electric utilities after deregulation. Wisconsin Electric had a model much closer to home: the natural gas industry. When deregulation took effect in 1985, major gas customers could buy directly from producers, paying local utilities like Wisconsin Natural Gas nothing more than a transportation fee. Competition blossomed, and the number of pipeline companies serving Wisconsin multiplied. Federal regulators went a large step further in 1992, stimulating competition by "unbundling" the gas industry into sales, transportation, and storage components. Customers, including local utilities, could mix and match vendors for maximum cost savings. The results mirrored those in the telecommunications field: greater complexity, more choices, and lower prices.

Deregulation was only one change affecting the system's gas subsidiary. In 1994, after months of negotiations, Wisconsin Energy bought Wisconsin Southern Gas and merged it into Wisconsin Natural. Based in Lake Geneva, Wisconsin Southern served an area that dovetailed perfectly with the southern border of WNG's territory; the acquisition also added 48,000 customers to the 288,000 already on Wisconsin Natural's lines. No sooner were the papers signed than Dick Abdoo announced a much larger move: the merger of Wisconsin Natural into Wisconsin Electric, effective January 1, 1996. Abdoo hoped that combining the companies would eliminate redundancies and therefore reduce costs — one of the primary goals of the revitalization effort. The move also marked a return of sorts: Not since the early days of TMER&L had all utility operations been combined in a single company.

As the revitalization campaign continued, on both the gas and the electric sides, Dick Abdoo tried to keep the process as open as possible. Rather than hatching a plan in secret and then imposing it on the work force, he took a "share as you go" approach, telling what he knew when he knew it. The predictable result was pervasive uncertainty; if even the chairman didn't have all the answers, then everything was up for grabs. Some employees considered the open-ended approach callous, even cruel, but it was completely intentional. Not only was corporate candor the most honest course, but it also helped to disabuse employees of their belief in entitlement. Only with the old culture erased, Abdoo believed, could a new, more progressive outlook emerge.

That outlook began to take shape on September 24, 1993. After months of research, discussion, and creative daydreaming, the revitalization team unveiled a brand-new structure for Wisconsin Electric. The company's departments had always been organized by function — legal, financial, construction, purchasing, and others — but "The Plan," as it was modestly called, placed the emphasis on *process*, defined by WEPCO's consultant as "a collection of activities that ... creates an output that is of value to the customer." The planning team identified five key processes: customer operations, fossil operations, nuclear power, bulk power, and a corporate center to support them all. The names were hardly ear-catching, but the activities they described became the core departments of a revitalized Wisconsin

Changes in the natural gas industry offered WEPCO a preview of what to expect in a deregulated world.

Electric. Each was conceived as a self-contained business unit, and each was expected to develop a structure based on service to customers, including internal customers. "With this plan," declared Dick Abdoo, "we have set in motion one of the most comprehensive and dramatic changes in our corporate history."

Implementation of the master plan tested the patience of even the most supportive employees. "There is certain to be some uncertainty and anxiety," Abdoo had warned, "until all of the pieces of the plan are in place." As literally thousands of workers changed titles, work stations, and responsibilities, life at Wisconsin Electric came to resemble a gigantic game of musical chairs. As the weeks passed, the tempo accelerated and the music got louder. A series of "quick hits" changed long-standing practices, and "outsourcing" claimed such time-honored functions as bill processing, stock transfer, and even, to a limited degree, engineering. The shifts had an obvious impact on the bottom line, but they were also intended to create a new kind of workplace. One of the master plan's stated goals was "to flatten the organization and push responsibility down," thereby helping to "empower employees and give them ownership of their work." Piece by piece, step by step, a substantially new structure and a substantially new outlook began to emerge.

The bad news was that the new Wisconsin Electric needed approximately 1,000 fewer people than the old. When the music stopped, nearly 20 percent of the work force would be left without seats. Local papers reported the magnitude of the planned cuts in December, 1993, a week before Christmas, and most WEPCO employees read the headlines in stunned silence. Working for the utility had long been the equivalent of academic tenure or civil-service employment: Once you were inside, you could reasonably expect to stay until retirement. As that expectation unraveled, some employees felt an acute sense of betrayal. After decades of operating under one set of rules, they were suddenly subject to another. More than a few, among them some senior executives, found the transition impossible to make.

Members of the revitalization team "invented" a new Wisconsin Electric.

Morale probably reached its lowest point in the first few months of 1994. As thousands pondered a wide range of early retirement and severance plans, not all of them voluntary, a chilling sense of insecurity gripped the workplace. Those who lost their jobs felt shell-shocked, and those who stayed behind felt survivor's guilt. Union leaders, representing nearly 70 percent of the work force, charged that management was focusing on customers and shareholders at the expense of its own employees. A series of focus groups conducted in April uncovered "a pervasive feeling of mistrust." In the meantime, so many workers took early retirement that the company began to hold mass open houses in the Public Service Building's auditorium to bid them all farewell.

Dick Abdoo was well aware of the trauma created by reengineering. As early as November, 1993, he acknowledged that the multiple changes were causing "considerable anxiety for many employees, as well as their families." The company encountered steady demand for a free booklet called *Adapting to Stress,* and the employee counseling program's caseload soared. Tensions were still high nearly a year after the master plan was announced. In the August, 1994, issue of *Synergy,* Abdoo made a succinct observation: "The transitional phase is hell."

Company veteran Dick Grigg succeeded Bill Boston as WEPCO's president in 1995.

The president in the power plant: The pressures of revitalization created an obvious need for closer ties between labor and management.

Despite the angst of his work force and the wear and tear on his own constitution, Dick Abdoo never wavered for a moment in his commitment to change. He and his team were guided by a firm belief that the organization's welfare was ultimately more important than the welfare of any individual, simply because the organization's failure would mean failure for every individual. "The good companies," Abdoo told the *Milwaukee Journal*, "won't wait until they're sick to take their medicine." Once WEPCO had chosen a course of treatment, the company's leader was determined to see it through to the end. "If we start, we finish," Abdoo told his planning team. "If we play, we play to win."

The players soon included the unions representing Wisconsin Electric employees. Revitalization put enormous pressures on the labor-management relationship, particularly in the early stages, and both sides struggled to find new grounds for understanding. The tangible result of their efforts was the Labor Management Committee, organized in September, 1994. Drawn equally from union leadership and WEPCO's executive ranks, the Committee began with a mutual acknowledgement that "security is ultimately dependent on market performance." The LMC's overriding goal has been to improve performance without destroying the "proper balance between technological and competitive realities and the needs and expectations of people." The company has willingly shared power in the effort to find that balance. The Labor Management Committee and its offshoots have been entrusted with difficult decisions on issues ranging from the use of temporary workers to the closing of facilities. Working together on these and other issues is beginning to build the bridge between labor and management, while providing a focus on success for both the company and employees.

It was not until the end of 1994, when the LMC was making its first decisions, that a semblance of calm returned to Wisconsin Electric. Systemwide employment had fallen by more than 800 people since 1993— a drop of nearly 14 percent — but the worst of the cuts were over. The new organizational structure was slowly taking shape, line by line, position by position, and new leaders were emerging. The most prominent was Dick Grigg, who became president of Wisconsin Electric on January 1, 1995. An engineer with broad experience in WEPCO's power plants, Grigg became the system's second-in-command when Bill Boston retired in mid-1995. Other promotions marked an emphasis on diversity. WEPCO's new vice-presidents included Calvin Baker (Finance), the company's first African-American executive officer, and three women: Kristine Krause (Fossil Operations), Kristine Rappé (Customer Services), and Ann Marie Brady (External Affairs). There was also unmistakable evidence of a youth movement. Between 1988, just before Abdoo's rise to power, and 1996, the average age of Wisconsin Electric's officers dropped from 53.6 to 46.

WEPCO's new executive team oversaw an enterprise that had changed drastically in a very short time. Some of the shifts shared an emphasis on economy. Numerous customer service centers were closed, meters were read every two months instead of every one, and nearly 1,000 employees were issued company credit cards to make small purchases. (The last move saved roughly 150,000 requisition forms a year.) Changes still in the offing will enhance the system's flexibility. "Broad bands" of compensation may replace narrow salary levels, a revised pension plan will increase both participation and "portability," and a novel job-sharing agreement already in place allows members of one union to work in another's jurisdiction during periods of seasonal slack. (The last change was suggested by union leaders.) Perhaps the most obvious transformation has involved the company's outlook and expectations. By 1995 Wisconsin Electric had become a leaner, more nimble enterprise, poised to take on the dragons that lurk in the uncharted wilderness of free competition.

Raising the Ante

No sooner had the dust begun to settle than Dick Abdoo stirred it up again, this time literally. In December, 1994, WEPCO announced plans to renovate the Public Service Building from top to bottom. The move was prompted, in part, by a desire to replace antiquated plumbing, electrical, and climate-control systems, but aesthetics played the larger role. Since 1906, when John Beggs dedicated the landmark, the PSB had suffered remodelings and "remuddlings" both large and small. The most objectionable was the fifth floor, tacked on in 1956 to provide space for an employee cafeteria and lounge. The addition was an architectural afterthought, perennially at odds with the classical lines of Beggs' monument, and it was slated for complete removal in 1996. In its place the architects substituted a cafeteria on the west side of the second floor, large enough to serve the reduced staffs in both the PSB and the Annex. Other elements of the multi-million-dollar project range from a restored auditorium to decorative carbarn "doors" on the outside walls — reminders of the PSB's heyday as an interurban train terminal. Construction crews launched their offensive on the west side of the building in March, 1995, gutting most of it by summer.

After a comprehensive remodeling, an office area on the second floor of the Public Service Building became the new employee cafeteria in 1996.

After months of dust, detours, and relocations, office workers gratefully observed completion of the first phase in early 1996.

The process of tearing down and rebuilding was not confined to the company's home office. The whole point of reengineering was to position WEPCO for success in a restructured utility industry, but the industry's future remained a tantalizingly open question. Hoping to influence and perhaps even accelerate change, Dick Abdoo and his colleagues took an active part in efforts to rebuild the electric utility business. Competition was a given. Federal authorities were dismantling the walls around monopolies, and Wisconsin's own Public Service Commission was beginning to redraw the lines on the playing field. In January, 1995, the PSC convened a multi-partisan advisory committee to hatch proposals for "a more market-oriented electric industry." Consensus proved elusive, but the Commission did not retreat from the basic concept of greater competition.

Dick Abdoo had put a team of planners to work on the restructuring problem long before the PSC committee convened. They started from a belief that competition, in and of itself, was a good thing, and that a lack of competition in the utility industry had created a dinosaur. Regulated monopolies, selling their energy on a cost-plus basis, had no particular incentives for excellence and no real disincentives for mediocrity; they controlled 100 percent of the market in either case. But anarchy was no alternative; the ruinous competition of the 1890s and the economic tyrannies of the 1930s served as chilling examples of what could develop in the absence of regulation.

WEPCO views the transmission system of the future as an open grid controlled by an independent operator.

Wisconsin Electric's vision of the future, unveiled in November, 1994, called for the best of both worlds. WEPCO's planners divided the electric utility business into three major processes: the generation, transportation, and sale of energy to end users. Why not, the planners asked, treat each process as a separate business activity, akin to manufacturing, distribution, and sales in other fields? Why not "unbundle" individual utilities and reconfigure the industry on the basis of core activities? A new group of generating companies, varying widely in price and reliability, would compete in the open market, just as automobile manufacturers vie for the same drivers. A host of customer service companies, each with its own products and personality, would compete for end users, just like automobile dealerships in local communities. The middle step — distribution — would remain a regulated monopoly, if only because competing networks of wires (or gas pipes) would be both unsightly and uneconomical. Generating companies would distribute their power through an independently managed transmission grid, a system that would serve as the true common carrier in its assigned region.

The gap between the present state of the American utility industry and the future envisioned for it by Wisconsin Electric may seem huge to a layperson, but WEPCO's leaders find it easy to bridge. Dick Abdoo, in particular, spends much of his time in the future, and his impatience for it to arrive is almost palpable. "The earlier the regulation changes," he said in 1994, "the better off we are." Why the excitement? For generations, electric utilities have conducted themselves as good little minnows in dull gray schools. The WEPCO plan would free them to grow colors, to swim in different schools, and to move together from a stagnant pond to the open sea.

Whether the waters of the future will bear much resemblance to those pictured by Dick Abdoo's team is anyone's guess, but the plan serves a useful purpose in the present. As debate over deregulation intensifies — in Washington, California, New England, and elsewhere — Wisconsin Electric is one of a handful of companies that can offer a coherent, well-reasoned view of an alternative future. The plan, therefore, is preemptive; it claims a position on the leading edge and dares others to follow. "We can't sit back and wait ... for someone else to set the rules and control the pace of change," said Abdoo. "Our plan puts Wisconsin on the map. It gives us a seat at the table."

One conclusion from the WEPCO plan is inescapable: The electric utility industry must consolidate. Regional service and regional competition cannot happen until the scores of electric fiefdoms and principalities combine into larger units. Wisconsin Electric has followed the advice implicit in its plan. On May 1, 1995, Dick Abdoo announced that Wisconsin Energy and Northern States Power planned to merge, forming a new enterprise called Primergy. Northern States is a Minneapolis-based utility serving an area that sprawls from Upper Michigan to the Dakotas, with more than 250,000 gas and electric customers in western Wisconsin. Abdoo and James Howard, NSP's chairman and chief executive, described the planned combination as a "merger of equals." Their agreement, pending regulatory approval, assures both companies an equal number of board seats, and it calls for Abdoo to assume the chairman's post when Howard retires in 2000.

Primergy would be incorporated as a Wisconsin company, but its headquarters would be in Minneapolis, the geographic center of the companies' combined service areas. The merger agreement would create two major subsidiaries: a new Wisconsin Energy Company, melding Wisconsin Electric with NSP's Wisconsin operations; and Northern States Power, consisting of the current NSP system in Minnesota and the Dakotas. Day-to-day operating decisions for the subsidiaries would be made in Milwaukee and Minneapolis, respectively. Primergy itself, with a market capitalization exceeding $6 billion, would be the tenth-largest energy company in the country.

In May, 1995, James Howard (left) and Dick Abdoo shocked observers by announcing plans to combine Wisconsin Electric and Northern States Power in a new enterprise: Primergy Corporation.

Announcement of the merger was a seismic event for both the utility industry and the communities served by the prospective partners. The shock waves are still being felt as the agreement makes its way through the regulatory maze. The Citizens' Utility Board (CUB), WEPCO's most persistent critic since 1979, has lambasted the proposed combination as a "Godzillatility." Some smaller energy companies have raised the specter of monopoly. More than one civic leader expresses fears that the merger would mean a loss of local control and local involvement to out-of-state interests. For Wisconsin Electric's own employees, the news was, in Yogi Berra's immortal phrase, a case of "déja vu all over again." Even before the company's reengineering was completed, they were being asked to start from scratch once more. Early estimates placed the cost savings of the merger at $2 billion over ten years, and it was no secret that a major portion of the savings would come from cutting redundant systems and employees. Stress levels began to rise again. "Take each day at a time," advised the editors of *Merger Update*, a new WEPCO publication. "Use your own stress management techniques, including exercise, talking with others and maintaining your sense of humor."

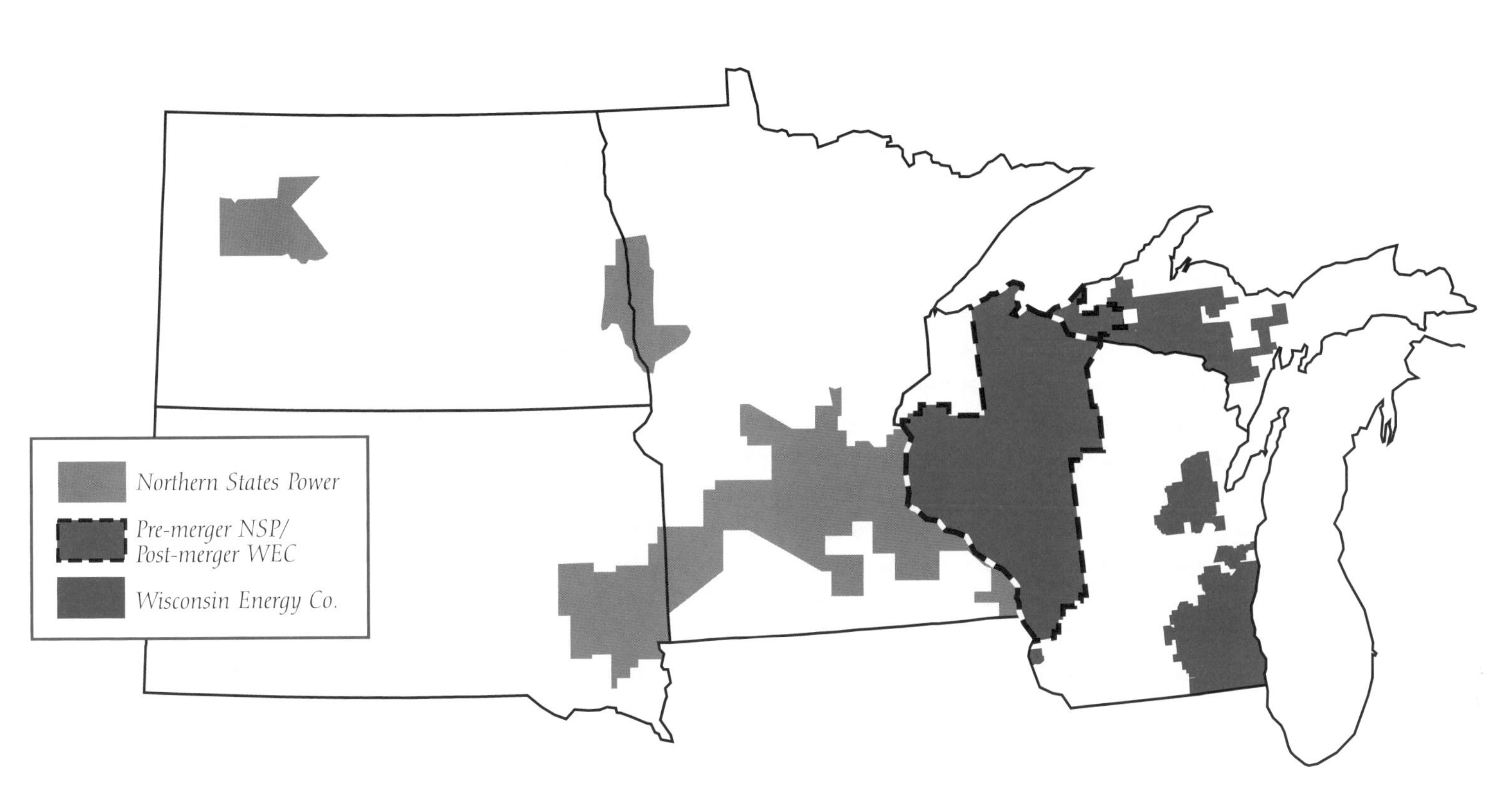

Primergy's proposed service area extends from Lake Michigan to the Dakotas, with all Wisconsin operations combined in a new Wisconsin Energy Company.

Whether he was addressing the fears of co-workers or the concerns of fellow business leaders, Dick Abdoo's response was always the same: Consolidation is inevitable. It is better to choose than be chosen, he argued, better to merge on your own terms than be swallowed by an unfriendly conglomerate. Joining a strong partner to form Primergy would allow WEPCO to run with a winner rather than languish at the back of the pack. "We have an opportunity," said Abdoo, "to build a brand-new company."

There are some unintended, perhaps unconscious, ironies in the pending merger. Until the Public Utility Holding Company Act of 1935 dissolved the system, Northern States Power was a unit of Standard Gas and Electric, a holding company formed by H.M. Byllesby. The Byllesby group's other holdings included Wisconsin Public Service Corporation, the same utility that still separates WEPCO's holdings in Upper Michigan, the Fox Valley, and southeastern Wisconsin. Wisconsin Electric now plans to take its neighbor's (and competitor's) place in the same historic system. The entire industry seems to be coming full-circle; Primergy bears more than a passing resemblance to the great regional utilities that flourished in the 1920s. The excesses of that era are not likely to recur, but the merger represents, in general terms, superpower revisited.

The proposed combination has not encountered a string of green lights, on either the state or the federal level. Whether it will become a reality on precisely the terms envisioned by the partners is, at this writing, less than certain. There is little doubt, however, that consolidation will play a major role in Wisconsin Electric's future. There is even less doubt that WEPCO will lead rather than follow the parade.

Whatever its final outcome, the planned merger capped a remarkable five years. Between 1990 and 1995, Dick Abdoo and his colleagues reengineered the company, reduced its work force, redirected its focus, and worked to restructure the American utility industry. Abdoo has tried, in a word, to *redo* practically everything. In the process, he has raised the ante time after time, moving from the company's internal culture to its structural underpinnings and finally to the foundations of the utility business. With the possible exception of John Beggs' early tenure, no five-year period in the utility's history has witnessed such sweeping changes.

There are, of course, risks involved.

By choosing to lead the industry's restructuring rather than follow, Wisconsin Electric risks getting so far in front of the parade that it takes the wrong route. "There is the chance," Abdoo concedes, "that you'll get disconnected." There is also the chance that restructuring will become a self-fulfilling prophecy. In its attempts to prepare for a particular future, WEPCO may help to precipitate precisely that future. That is a risk that Abdoo is entirely willing to accept, even embrace. It is better, in his view, to anticipate the future than to live in the past, even if it means that you move too fast on occasion.

Another risk involves the Wisconsin Electric work force. The system's employees have operated under consistently high levels of stress for several years; they have weathered job insecurities, deep personnel cuts, multiple reassignments, periodic physical relocations, and the anxieties arising from the proposed merger with Northern States Power. Frayed nerves and poor concentration might be expected, but the people of Wisconsin Electric have so far shown impressive resilience. Not only have the numerous shake-ups been largely invisible to the public, but WEPCO's employees have risen to unusual challenges, particularly in the torrid summer of 1995. As a series of heat waves pushed peak demand to record levels, utility crews were kept busy hosing down overworked transformers and responding to heat-related outages. Conditions neared the crisis stage on July 14, when a generator failure at Point Beach reduced the system's capacity by a whopping 500 megawatts. Only a coordinated plea for reduced consumption and a quick response from customers averted widespread outages. On August 28, just six weeks later, the most severe thunderstorm in several years rumbled through eastern Wisconsin, leaving 100,000 customers without power. Working feverishly to clear fallen trees and repair downed lines, trouble crews restored service with a minimum of inconvenience. Dick Grigg, WEPCO's new president, praised the work ethic evident during the multiple emergencies. "In these times of continuing change and high anxiety," he wrote, "it shows once again that the basic fabric of people at Wisconsin Electric is still one of our key strengths."

After a Century

In years to come, Richard A. Abdoo may be hailed as a seer, a prophet, a Moses who led his company to the Promised Land of free competition and runaway prosperity. He may also be remembered as the leader who moved too far and too fast for prevailing conditions. It is much too early to tell which viewpoint will prevail, but Abdoo has already earned a place as one of the leading change agents in Wisconsin Electric's history. Anticipating what he views as overwhelming market forces, the system's chief executive has recast the assumptions underlying his company's role in its industry. Abdoo has

The faces and the fashions have changed enormously in the last century ...

transformed WEPCO, at least in prospect, from a regulated monopoly to an aggressive competitor, and he has led his firm to the brink of a merger that would create the dominant energy company in the Upper Midwest.

His initiatives ensured a memorable centennial year for Wisconsin Electric. Abdoo designated 1996 as the year of transition, the dividing line between before and after. In 1997, if all goes according to plan, WEPCO's internal retooling will be finished, and the shape of the restructured utility industry will be apparent, at least in rough outline. In a much more visible transition, Wisconsin Electric is slated to go out of existence, subsumed in a new and larger Wisconsin Energy.

As WEPCO pauses at the century milestone, the company is without question a work in progress. The final coats of paint aren't even on yet, much less dry. Some traditionalists have already decided that they don't like the new colors. They consider the cumulative changes of the Abdoo years a painful discontinuity, a sharp break with a proud past. What they may fail to consider is the fact that Wisconsin Electric has always been a work in progress. Its history is a chronicle of constant change, and its oldest cultural trait is adaptability. The enterprise emerged from competitive chaos in the 1890s, brought to life by a pair of Henrys, Villard and Payne, who saw opportunity in disorder — not unlike Richard Abdoo. Their creation, under John Beggs, became a superlative streetcar company, one that saw the light — Thomas Edison's light — only gradually. James Mortimer reordered its priorities, shifting the utility's emphasis from transit to power and light. When the electric industry consolidated in the 1920s, TMER&L became, under North American's banner, part of a superpower empire that served a million customers in ten states. When automobile competition proved insurmountable in the 1930s, TM severed its transit arm and changed its name to Wisconsin Electric. When World War II and the boom that followed pushed the demand for power to new heights, WEPCO ran to keep pace. When conservation became imperative in the 1970s, the utility intentionally slowed down.

... but the people of Wisconsin Electric continue to keep the power flowing.

The multiple shifts and adaptations have been much more than reactions. Through all the turning points of its first century, Wisconsin Electric has displayed an unmistakable pioneering spirit. It was, for starters, the first integrated utility in the nation, the first to combine traction, light, and power in a single regional system. Once established, the company displayed a strong contrarian streak, a definite inclination to take the road less-traveled, the path overlooked or ignored by its counterparts. The result is a record of innovations that no other company can claim: the masterful expansion campaigns of John I. Beggs, the unprecedented employee programs of the Mortimer era, the birth of pulverized-coal technology in 1919, the all-out marketing programs of the Depression era, and the demand-side initiatives of the McNeer years. The new directions plotted by Dick Abdoo are not without precedent. WEPCO's current chairman is building on an old tradition to create what he calls "a new yesteryear," a record of achievement that employees can be proud of a century from now.

What that century will bring is anyone's guess. More change, certainly; Wisconsin Electric has compressed decades of evolution into a few years, and the velocity of change is likely to increase. "I don't foresee a point," said Abdoo, "at which we can say, 'OK, now we're finished.'" But a core constant still applies: From whatever source and for whatever purpose, WEPCO is an indispensable provider of energy in an area stretching from Upper Michigan to the Illinois border. The company has generated the region's future, and it has earned, along the way, a reputation for innovation, a continuing place on the frontiers of the possible. In the decades to come, the company will operate under different leaders, even under a different name, but the enterprise will always follow the path its founders chose in 1896 — the path of a pioneer.

Primary Corporate Sources

Annual Reports:

The Milwaukee Electric Railway and Light Co. (1913-1937)
Wisconsin Electric Power Co. (1938-1985)
Wisconsin Energy Corp. (1986-1995)
Wisconsin Gas and Electric Co. (1926-1949)
Wisconsin Michigan Power Co. (1929-1965)

Corporate Minutes:

Milwaukee Street Railway (1895-1896)
The Milwaukee Electric Railway and Light Co. (1896-1938)
The Milwaukee Light, Heat and Traction Co. (1899-1919)
Wisconsin Electric Power Co. (1938-1995)
Wisconsin Energy Corp. (1987-1995)
Wisconsin Gas & Electric Co. (1931-1949)
Wisconsin Michigan Power Co. (1928-1942)

Employee Periodicals:

Rail & Wire (1913-1953)
The Outlet (1954-1987)
Currently (1980-1995)
Synergy (1989-1995)
Revitalization Update (1993-1994)
Merger Update (1995)

Sources of General Interest

Abendroth, H.G. "Historical Facts of Milwaukee's Street Railway System." Milwaukee: unpublished manuscript, 1915.
Canfield, Joseph M., ed. *TM: The Milwaukee Electric Railway and Light Company.* Chicago: Central Electric Railfans Association, 1972.
Canfield, Joseph M., ed. *Badger Traction.* Chicago: Central Electric Railfans Association, 1969.
Electric Railway Journal, Electrical World, Street Railway Journal, Street Railway Review, and *Transit Journal.* Various issues, 1896-1944 (fully indexed).
The Electric Railways of Wisconsin. Chicago: Central Electric Railfans Association, 1953.
Hyman, Leonard S. *America's Electric Utilities: Past, Present and Future.* Arlington, VA: Public Utilities Reports, 1988.
Mayer, Henry M. *A Backward Glance: The Street Railway in Milwaukee.* Milwaukee: undated monograph, c. 1970.
McDonald, Forrest. *Let There Be Light: The Electric Utility Industry in Wisconsin, 1881-1955.* Madison: American History Research Center, 1957.
McShane, Clay. *Technology and Reform: Street Railways and* the *Growth of Milwaukee, 1887-1900.* Madison: State Historical Society of Wisconsin, 1974.
Middleton, William D. The *Interurban Era.* Milwaukee: Kalmbach Publishing Co., 1961.
Milwaukee Journal and *Milwaukee Sentinel* clippings, "Wisconsin Electric Co.," 1940-1995. Milwaukee Public Library.
Nye, David E. *Electrifying America: Social Meanings of a New Technology.* Cambridge: MIT Press, 1990.
Schultz, Russell E. *A Milwaukee Transport Era.* Glendale, CA: Interurbans, 1980.
Wisconsin Utilities Association. Roundtable interview with state utility leaders, May 26, 1954. State Historical Society collection.

Selected Chapter Sources

Chapters 1 and 2 (1880-1896)

Brewer, Thomas, ed. *The Robber Barons: Saints or Sinners?* New York: Holt, Rinehart & Winston, 1970.
Canby, Edward T. *A History of Electricity.* New York: Hawthorn Books, 1963.
Carlin, Kathleen M. "'Boss' John A. Hinsey." *Historical Messenger of the Milwaukee County Historical Society.* June, 1962.
"The Cream City." Milwaukee: *Evening Wisconsin* souvenir edition, 1891.
Hammond, John W. *Men and Volts: The Story of General Electric.* Philadelphia: J.B. Lippincott Publishers, 1941.
Kellogg, Louise P. "The Electric Light System at Appleton." *Wisconsin Magazine of History.* December, 1922.
McDonald, Forrest. "The People Get a Light." *Wisconsin Magazine of History.* Summer, 1954.
Meyer, Herbert W. *A History of Electricity and Magnetism.* Cambridge: MIT Press, 1971.
Milwaukee Sentinel. Various articles, 1859-1891, indexed under "Street Railways," "Electric...," "Appleton," "Henry C. Payne," "John Hinsey," etc. in WPA index, Milwaukee Public Library.
Rail and Wire. "Early Days of Electric Lighting in Milwaukee." July-November, 1938.
Rail and Wire. Fiftieth-anniversary issue, January, 1946.
Ryan, Thomas, ed. *History of Outagamie County.* Chicago: Goodspeed Historical Association, 1911.
... they turned on the lights. Appleton: Wisconsin Michigan Power Co., 1954.
Villard, Henry. *Memoirs of Henry Villard: Journalist and Financier, 1835-1900.* Boston: Houghton, Mifflin Co., 1904.
Wight, W.W. *Henry Clay Payne: A Life.* Milwaukee, 1907.

Chapter 3 (1896-1911)

Beggs, John I. Testimony before Railroad Commission in City of Milwaukee rate case. Mar. 19, 1907.
Butler, John A. "The Street Railway Problem in Milwaukee." *Municipal Affairs.* March, 1900.
Damaske, Charles H. *Along the Right-of-Way to East Troy.* Self-published, 1989.
Dean, Arthur H. *William Nelson Cromwell, 1854-1948.* Self-published, 1957.
East Troy News. Various articles, 1903-1911.
Gavett, Thomas. *Development of the Labor Movement in Milwaukee.* Madison: University of Wisconsin Press, 1965.
In Memoriam: John I. Beggs, 1847-1925. Milwaukee, c. 1925.
Lush, Charles K. *The Autocrats.* New York: Doubleday, Page & Co., 1901.
McDonald, Forrest. "Milwaukee Streetcars and Politics, 1896-1901." *Wisconsin Magazine of History.* Spring, 1956.
Mueller, Donald J. *Notes on the Great Milwaukee Strike of 1896.* Milwaukee: University of Wisconsin-Milwaukee, 1978.
Ondercin, David C. "Corruption, Conspiracy, and Reform in Milwaukee, 1901-1909." *Historical Messenger of the Milwaukee County Historical Society.* December, 1970.
Ranney, Joseph A. "The Political Campaigns of Mayor David S. Rose." *Milwaukee History.* Spring, 1981.
Rau, O.M., comp. *Scenic Trolley Rides.* Milwaukee: TMER&L, 1898.
Thelen, David P. *The New Citizenship: Origins of Progressivism in Wisconsin, 1885-1900.* Columbia, MO: University of Missouri Press, 1972.
Wieland, Lauretta L., ed. *The Pewaukee Lake Playground.* Self-published, 1981.
Young, Andrew D. and Provenzo, Eugene F., Jr. *The History of the St. Louis Car Company.* San Diego: Howell-North Books, 1978.

Chapters 4 and 5 (1911-1929)

America's New Frontier. Chicago: Middle West Utilities, 1929.
Hoan, Daniel. *The Failure of Regulation.* Chicago: Socialist Party of the United States, 1914.
Our Golden Years. Milwaukee: Schuster's Department Store, 1934.
Our Songs of Labor. Milwaukee: TMER&L, 1916.
St. Clair, David. *The Motorization of American Cities.* New York: Praeger Publishers, 1986.
Utility Corporations Reports. U.S. Senate Document 92: 33-34, vol. 14, 1931. (FTC investigation of North American Co.)
Wisconsin Highways, 1835-1945. Madison: State Highway Commission, 1947.

Chapter 6 (1930-1945)

Biemiller, Andrew and Hannah. "General Strike — New Style." *The World Tomorrow.* July 12, 1934.
Higgins, Florence. "Trial by Fire." *The Nation.* July 18, 1934.
Milwaukee Journal and *Milwaukee Sentinel.* Strike articles, June 20-July 1, 1934.
"President Roosevelt Stresses the Need for Holding Company Legislation." *Congressional Digest.* May, 1935.
Ramsay, Marion L. *Pyramids of Power.* Indianapolis: Bobbs-Merrill Co., 1937.
Report on Municipal Light and Power System, Milwaukee. Burns & McDonnell Engineering Co., 1934.

Chapters 7–10 (1945-1995)

Alexanderson, E. Pauline, ed. *Fermi-1: New Age for Nuclear Power.* La Grange Park, IL: American Nuclear Society, 1979.
Clarfield, Gerard H. and Wiecek, William M. *Nuclear America.* New York: Harper & Row, 1984.
Kaku, Michio and Trainer, Jennifer, eds. *Nuclear Power: Both Sides.* New York: W.W. Norton and Co., 1982.
Sheffer, Victor B. *The Shaping of Environmentalism in America.* Seattle: University of Washington Press, 1991.

Interviews

Richard Abdoo	Charles McNeer
Russell Britt	Wilfred Pollock
Sol Burstein	Norbert Schuh
Alfred Gruhl	Ezra Sorenson
Robert Kloes	Lawrence Wangerin
Richard Leach	Howard Warhanek

Assistance with transit illustrations was graciously provided by railfans Jack Galloway, Don Leistikow, Duane Matuszak, Bill Nedden, Russ Schultz, and Ed Wilkommen.

Special thanks to John Speaker and Ann Marie Brady for oversight, John Bartel for project management, Christine Kowalski for graphic design, and Darlene Waterstreet for archival assistance.

Index

Note: Numbers in italics refer to illustrations.

John Gurda is a Milwaukee-based writer and historian. He is the author of 11 previous books, on subjects ranging from Frank Lloyd Wright to life insurance and from ethnic neighborhoods to steel foundries. Gurda is a six-time winner of the Award of Merit from the State Historical Society of Wisconsin. He is currently working on a general history of the Milwaukee area.

organized on the 29th inst.,
nsin, by the purchasers abo
e Milwaukee Electric Rail
uant to Section 1788 of the
other statutes of Wiscons
were presented to the
n motion, it was duly a
Resolved, that the
Milwaukee Electric Railwa
under the laws of the Stat